Oligonucleotide synthesis

a practical approach

Edited by
M J Gait

MRC Laboratory of Molecular Biology,
Hills Road, Cambridge CB2 2QH, England

This book belongs to

_____

_____

_____

DISTRIBUTED BY

 ALDRICH CHEMICAL CO.

P.O. Box 355, Milwaukee, Wisconsin 53201 USA

Telephone	414-273-3850
TWX	910-262-3052 Aldrichem MI
Telex	26 843 Aldrich MI
FAX	414-273-4979

To order
Toll-free USA/Canada   800-588-9160
P.O. Box 2060   Milwaukee, Wisconsin 53201   USA

IRL PRESS
Oxford · Washington DC

IRL Press Limited,
P.O. Box 1,
Eynsham,
Oxford OX8 1JJ,
England

© 1984 IRL Press Limited

All rights reserved by the publisher. No part of this book may be reproduced or transmitted in any form by any means, electronic or mechanical, including photocopying, recording or any information storage and retrieval system, without permission in writing from the publisher.

First published October 1984
Revised reprinting July 1985

British Library Cataloguing in Publication Data

Oligonucleotide synthesis : a practical approach.—(Practical approach series).
  1. Nucleic acid synthesis
I. Gait,M.J.   II. Series.
574.7′90459   QD433

ISBN 0-904147-74-6

Cover illustration. The design for the cover was based on Figure 5 from Chapter 1, showing the four common heterocyclic bases and numbering system for the primary structure of DNA.

Printed in England by Information Printing, Oxford.

Preface

You might wonder what a chemistry book is doing in a series of practical handbooks designed primarily for workers in the biological sciences. Yet anyone who has witnessed the increasingly common sight of a biologist ruefully peering over a tiny glass funnel full of silica gel, or a biochemist tinkering in vain with his brand new DNA synthesis machine, will know immediately why this book has been written. Oligonucleotide synthesis has become an essential technique required by many biologically orientated laboratories. Yet up to now there has been no practical manual available that adequately covers all the important techniques of synthetic oligonucleotide chemistry in a style suitable to be followed by a non-specialist. This book aims to fill that gap.

In order to reduce to manageable proportions a subject which has undergone a dramatic expansion in recent years, I have selected only those techniques central to the task of oligonucleotide synthesis. Since synthetic DNA is currently in much greater demand than RNA, DNA synthesis forms the predominant part of the book. Here I have chosen just two chemical methods for assembly of oligodeoxyribonucleotide chains: phosphite-triester (Chapter 3) and phosphotriester (Chapter 4) which involve the now established technique of solid-phase synthesis. These two methods are widely and successfully used in laboratories worldwide and can therefore be thoroughly recommended. Other chapters describe the important techniques of preparation of protected deoxyribonucleoside starting materials (Chapter 2), purification of oligonucleotides by high-performance liquid chromatography (Chapter 5) and purification and sequence determination of synthetic oligonucleotides (Chapter 6). I have also taken the unusual step for books in this series by adding a brief introduction to methods of DNA synthesis, written nevertheless from a practical angle (Chapter 1).

In Chapter 7 the methods described for chemical synthesis of oligoribonucleotides reflect the more conventional techniques of chemistry in solution appropriate for the often larger quantities of synthetic RNA required. To complement this, Chapter 8 describes the important role that enzymes now play in preparation of longer RNA fragments.

My co-authors in this project come with the highest possible credentials. Each of the research groups is highly active in the development of methods of oligonucleotide synthesis and many of the authors have instructed in laboratory courses designed for non-specialists. The manuscripts submitted were of very high standard and as a result only relatively few alterations were needed, and this has made my role as editor an unusual pleasure. The enthusiasm of all parties concerned, including the publishers, has resulted in the remarkable achievement of publication in under a year from original conception.

M.J.Gait

Acknowledgement

The editor would like to thank the many members of the MRC Laboratory of Molecular Biology for their help with various aspects of this book. Special thanks are due to Joan Illsley for invaluable secretarial assistance.

Contributors

Tom Atkinson
Department of Biochemistry, University of British Columbia, 2146 Health Sciences Mall, Vancouver, British Columbia, Canada V6T 1W5

Dorothy Beckett
Department of Biochemistry, 415 Roger Adams Laboratory, University of Illinois, 1209 West California Street, Urbana, IL 61801, USA

Michael J.Gait
MRC Laboratory of Molecular Biology, Hills Road, Cambridge CB2 2QH, UK

Fawzy Georges
Section of Biochemistry, Molecular and Cell Biology, Wing Hall, Cornell University, Ithaca, NY 14853, USA

Zahre Hanna
Section of Biochemistry, Molecular and Cell Biology, Wing Hall, Cornell University, Ithaca, NY 14853, USA

Roger A.Jones
Department of Chemistry, Rutgers University, Douglass College, New Brunswick, NJ 08903, USA

Larry W.McLaughlin
Max-Planck-Institut für Experimentelle Medizin, Abteilung Chemie, Hermann-Rein-Strasse 3, D-3400 Göttingen, FRG

Saran Narang
Molecular Genetics Section, Division of Biological Sciences, National Research Council of Canada, Ottawa, Ontario, Canada K1A OR6

Norbert Piel
Max-Planck-Institut für Experimentelle Medizin, Abteilung Chemie, Hermann-Rein-Strasse 3, D-3400 Göttingen, FRG

Michael Smith
Department of Biochemistry, University of British Columbia, 2146 Health Sciences Mall, Vancouver, British Columbia, Canada V6T 1W5

Brian S.Sproat
MRC Laboratory of Molecular Biology, Hills Road, Cambridge CB2 2QH, UK. Present address: EMBL, Meyerhofer Strasse 1, D-6900, FRG

Olke C.Uhlenbeck
Department of Biochemistry, 415 Roger Adams Laboratory, University of Illinois, 1209 West California Street, Urbana, IL 61801, USA

Jacques H. van Boom
Gorlaeus Laboratories, Department of Chemistry, State University of Leiden, P.O.Box 9502, 2300 RA Leiden, The Netherlands

Carl T.J.Wreesman
Gorlaeus Laboratories, Department of Chemistry, State University of Leiden, P.O.Box 9502, 2300 RA Leiden, The Netherlands

Nai-Hu Wu
Section of Biochemistry, Molecular and Cell Biology, Wing Hall, Cornell University, Ithaca, NY 14853, USA

Ray Wu
Section of Biochemistry, Molecular and Cell Biology, Wing Hall, Cornell University, Ithaca, NY 14853, USA

Contents

ABBREVIATIONS	xiii
HEALTH WARNING	xiii

1. AN INTRODUCTION TO MODERN METHODS OF DNA SYNTHESIS   1
Michael J.Gait

Introduction	1
The Need for Synthetic Oligodeoxyribonucleotides	1
Construction of duplex DNA	1
Oligodeoxyribonucleotides as primers	3
Selection of recombinant DNA	5
Site-directed mutagenesis	6
Structural studies involving synthetic DNA	7
DNA Structure and Chemical Reactivity	8
DNA primary and secondary structure	8
Chemical reactivity of DNA	10
Chemical Synthesis of Oligodeoxyribonucleotides	11
Protecting groups	12
The solid-phase method	14
Pitfalls for the Unwary: Practical Advice to the Novice	18
Solvents and reagents	18
Apparatus	19
Techniques and trouble-shooting	21
References	21
Update	22

2. PREPARATION OF PROTECTED DEOXYRIBONUCLEOSIDES   23
Roger A.Jones

Introduction	23
Per-Acylation	24
General procedure for N-acylation by per-acylation	24
Transient Protection	25
General procedure for N-acylation by transient protection	25
6-N,N-Dibenzoyl-2'-deoxyadenosine	27
Protection of the 5'-Hydroxyl Group as a 4,4'-Dimethoxytrityl Ether	27

vii

Reaction of N-acyl deoxyribonucleosides and thymidine with 4,4'-dimethoxytrityl chloride	27
One-flask $O^{5'}$ and N protection of deoxyadenosine and deoxycytidine	28
O^6 Protection of Deoxyguanosine	29
Synthesis of O^6-nitrophenylethyl or O^6-cyanoethyl deoxyguanosine derivatives by sulphonylation/displacement	31
Synthesis of O^6-nitrophenylethyl deoxyguanosine by Mitsunobu alkylation	32
Molecular Weights and Volumes	33
References	34

3. SOLID PHASE SYNTHESIS OF OLIGODEOXYRIBONUCLEOTIDES BY THE PHOSPHITE-TRIESTER METHOD 35
Tom Atkinson and Michael Smith

Introduction	35
The Chemistry of Solid-Phase Phosphite-Triester Synthesis	35
Synthesis of Protected Deoxyribonucleoside-3'-O- Diisopropyl Phosphoramidites	39
Preparation of methyl phosphodichloridite	39
Preparation of chloro-N,N-diisopropylamino methoxyphosphine	41
Preparation of deoxyribonucleoside-3'-O(N,N-diisopropylamino)phosphoramidites	41
Preparation of Solid Supports	45
Preparation of aminopropyl-derivatised Fractosil 500	46
Preparation of deoxyribonucleoside-3'-O-succinates and the deoxyribonucleoside-derivatised silica gels	47
Controlled pore glass solid support	49
Assembly of Protected Oligodeoxyribonucleotides	49
Reactions involved in the assembly cycle	49
Reagents and solvents	51
Apparatus	54
Procedure for oligodeoxyribonucleotide synthesis	58
Preparation for synthesis	64
Synthesis cycle	65
Removal of Oligodeoxyribonucleotide from Solid Support, Deprotection and Isolation	67
Deprotection procedure	68
Alternative deprotection procedure	69
Purification of oligonucleotides by gel electrophoresis	70
Scale Up or Down, Multiple Parallel Synthesis and Synthesis of Mixed Sequence Oligodeoxyribonucleotides	72

Scale up or down	72
Multiple simultaneous synthesis	73
Synthesis of mixed sequence probes	75
Nucleoside-specific coloured trityl cations	75
Variations in procedures	76
Chemicals and Equipment	78
Oligodeoxyribonucleotide synthesis equipment	78
Oligodeoxyribonucleotide deprotection	78
Oligodeoxyribonucleotide purification by gel electrophoresis	78
Reagents for oligodeoxyribonucleotide synthesis	79
Solvents for oligodeoxyribonucleotide synthesis	79
Reagents for preparing deoxyribonucleoside-3'-O-phosphoramidites	80
Reagents for preparing deoxyribonucleoside-derivatised supports	80
References	81
Update	81

4. SOLID-PHASE SYNTHESIS OF OLIGODEOXYRIBO-NUCLEOTIDES BY THE PHOSPHOTRIESTER METHOD 83
Brian S.Sproat and Michael J.Gait

Introduction	83
General Methods and Precautions	83
Sources of potential trouble	83
The support	84
Protecting groups	86
Chain assembly	87
Deprotection and purification	89
Experimental Procedures	89
Preparation of supports	89
Preparation of deoxyribonucleotide monomers, deoxyribonucleoside-3'-O-succinates and coupling agent	91
Purification of solvents and reagents	98
Synthesis of Fully Protected Oligodeoxyribonucleotides	101
Setting up a manual synthesis apparatus	101
Preparing the materials for a small-scale synthesis on controlled pore glass	103
Assembly of the fully protected support-bound oligodeoxyribonucleotide on small scale on CPG	104
Large-scale synthesis of oligodeoxyribonucleotides	107
Cleavage of the Oligodeoxyribonucleotides from the Support and Subsequent Deprotection	108
Purification Procedures	109
Ion-exchange h.p.l.c.	109

Desalting	109
Reversed phase h.p.l.c.	110
Chemicals and Equipment	110
Supports	110
Succinates	111
2'-Deoxyribonucleosides	111
5'-O-Pixyl protected 2'-deoxyribonucleosides	111
5'-O-Dimethoxytrityl protected monomers	111
Solvents	111
Miscellaneous reagents	112
Special equipment	113
T.l.c. plates	114
Parts list for dual column Omnifit synthesiser	114
References	114
Update	115

5. CHROMATOGRAPHIC PURIFICIATION OF SYNTHETIC OLIGONUCLEOTIDES   117
Larry W.McLaughlin and Norbert Piel

Introduction	117
Equipment for High Performance Liquid Chromatography	117
The Chromatographic Support	119
Bonded phase silica supports	119
Pre-packed or self-packed columns	119
Selected Applications	120
Separation of synthetic oligodeoxyribonucleotide mixtures	120
Isolation of oligodeoxyribonucleotide products	123
Analysis of oligodeoxyribonucleotide products	126
Isolation of oligoribonucleotides	127
Analysis of oligoribonucleotides	128
Isolation and analysis of oligoribonucleotides produced by enzymatic synthesis	129
Mixed Mode Chromatography	131
Conclusions	132
References	132

6. PURIFICATION AND SEQUENCE ANALYSIS OF SYNTHETIC OLIGODEOXYRIBONUCLEOTIDES   135
Ray Wu, Nai-Hu Wu, Zahre Hanna, Fawzy Georges and Saran Narang

Introduction	135

Purification Procedures	137
T.l.c. on silica gel plates	137
Polyacrylamide slab gel electrophoresis	138
Methods for Labelling Oligodeoxyribonucleotides	139
Labelling of 5′ ends of oligodeoxyribonucleotides with T4 polynucleotide kinase and [γ-^{32}P]ATP	139
Labelling of 3′ ends of oligodeoxyribonucleotides with terminal transferase and [α-^{32}P]cordycepin-5′-triphosphate	140
Sequence Determination by Enzymatic Degradation and Two-Dimensional Chromatography	141
Sequence Determination by Chemical Cleavage Reactions and Gel Electrophoresis	143
Base-specific chemical cleavage reactions	143
Sequencing gels	147
References	151

7. CHEMICAL SYNTHESIS OF SMALL OLIGORIBONUCLEOTIDES IN SOLUTION 153
Jacques H.van Boom and Carl T.J.Wreesman

Introduction	153
Synthesis of 2′-O-Tetrahydropyranylribonucleoside Monomers	156
Synthesis of 2′-O-tetrahydropyranyluridine (**1a**)	156
Synthesis of 2′-O-tetrahydropyranyl-4-N-anisoylcytidine (**1b**)	158
Synthesis of 2′-O-tetrahydropyranyl-6-N-benzoyladenosine (**1c**)	162
Synthesis of 2′-O-tetrahydropyranyl-2-N-diphenylacetylguanosine (**1d**)	165
Synthesis of 2′-O-tetrahydropyranyl-2-N-acetyl-6-O-(4-nitrophenylethyl)guanosine (**1e**)	168
Synthesis of 2′-O-Tetrahydropyranyl-5′-O-(2-Hexadecyloxy-4′,4″-dimethoxytrityl)ribonucleoside Monomers	171
Reagents and solvents	171
Synthesis of 2-hexadecyloxy-4′,4″-dimethoxytrityl alcohol	172
Synthesis of the ribonucleoside monomers	173
Synthesis of a Fully Protected Short Oligoribonucleotide	174
Reagents and solvents	174
Synthesis of phosphorylating agent: 2-chlorophenyl-O,O-bis[1-benzotriazolyl]phosphate (**3**)	174
Synthesis of dimer (**5**)	175
Synthesis of trimer (**6**)	176
Synthesis of tetramer (**7**)	176
Synthesis of pentamer (**8**)	176
Synthesis of hexamer (**9a**)	176

Deblocking of the Fully Protected Hexaribonucleotide **9a**	177
Reagents and solvents	
Removal of the 2-hexadecyloxy-4′,4″-dimethoxytrityl group from hexamer **9a**	177
Acetylation of the free hydroxyl group of hexamer **9b**	178
Removal of the base-labile groups from hexamer **9c**	178
Purification of the partially deblocked hexamer **9d**	178
Removal of the 2′-O-tetrahydropyranyl groups from hexamer **9d**	179
The Preparation of Longer RNA Fragments	179
Synthesis of a hexadecamer RNA fragment	179
Future trends	182
Acknowledgements	182
References	182

8. ENZYMATIC SYNTHESIS OF OLIGORIBONUCLEOTIDES   185
Dorothy Beckett and Olke C. Uhlenbeck

Introduction	185
General Strategy	186
Synthesis of short blocks	186
Joining of blocks with RNA ligase	188
Handling RNA for Enzymatic Synthesis	190
Equipment	190
Buffers and supplies	190
Separation and purification of oligonucleotides	191
Polynucleotide Phosphorylase Reactions	192
Materials	192
Equilibrium polynucleotide phosphorylase reaction	193
Nuclease-assisted reactions	194
Polynucleotide Kinase and RNA Ligase Reactions	194
Materials	194
Polynucleotide kinase reaction	195
RNA ligase reactions	195
References	196

APPENDICES

I. General Laboratory Techniques for Oligonucleotide Synthesis	199
II. H.p.l.c. Column Packing Techniques	207
III. Suppliers of Chemicals and Equipment for Oligonucleotide Synthesis	211
INDEX	**215**

Abbreviations

APS	aminopropylsilyl
CNE	cyanoethyl
CPG	controlled pore glass
DBU	1,8 diazabicyclo[5.4.0]undec-7-ene
DCA	dichloroacetic acid
DCCI	N,N'-dicyclohexylcarbodiimide
DCE	1,2-dichloroethane
DCU	dicyclohexylurea
DMAP	4-dimethylaminopyridine
DMF	N,N-dimethylformamide
DMTr	dimethoxytrityl (4,4'-dimethoxytriphenylmethyl)
DTT	dithiothreitol
Hepes	N-2-hydroxyethylpiperazine-N'-2-ethanesulphonic acid
MMTr	monomethoxytrityl
MSNT	1-mesitylenesulphonyl-3-nitro-1,2,4-triazole
NPE	nitrophenylethyl
ODS	octadecylsilyl
PEI	polyethyleneimine
SAX	strong anion-exchanger
SDS	sodium dodecyl sulphate
TCA	trichloroacetic acid
TEA	triethylamine
THF	tetrahydrofuran
THP	tetrahydropyranyl
TMED	N,N,N',N'-tetramethylethylenediamine
TMS	tetramethylsilane

HEALTH WARNING
USE OF CHEMICAL REAGENTS

Many of the chemicals used in oligonucleotide synthesis are toxic. Some are corrosive and others flammable. Wherever possible in this book specifically known hazards are highlighted. In general it is important when handling organic chemicals to abide by good safety standards, i.e., use of a laboratory coat, safety spectacles and disposable gloves. Particular care should be taken when carrying out procedures such as distillation, use of vacuum lines or bottled gases. The use of a fume hood is highly recommended.

CHAPTER 1

An Introduction to Modern Methods of DNA Synthesis

MICHAEL J. GAIT

1. INTRODUCTION

In 1967 when Khorana announced his intention to synthesise a gene chemically (1) and then with an army of co-workers proceeded to carry out his promise, there were some biologists who questioned whether such synthetic DNA would ever have any practical application. The numerous uses developed for synthetic DNA in recent years have dispelled all such doubts, and the ability to synthesise DNA fragments rapidly has become an important asset in many laboratories. Substantial advances in techniques have now put synthesis of oligodeoxyribonucleotides within the compass of the non-specialist, but with a bewildering selection of alternative methods available it is not always easy to choose the ones most appropriate for the particular application sought. This chapter outlines the most common applications and their synthetic requirements, gives an overview of the basic strategies for chemical synthesis of oligodeoxyribonucleotides and finally comments on some of the practical decisions needed to be taken by those contemplating setting up a DNA synthesis facility.

2. THE NEED FOR SYNTHETIC OLIGODEOXYRIBONUCLEOTIDES

The majority of applications of synthetic oligodeoxyribonucleotides are in the area of recombinant DNA (2). In particular they are used for construction, selection and determination of the DNA sequence of recombinants as well as the more recent application of site-directed mutagenesis. The techniques of cloning enable single recombinant molecules of DNA to be selected and amplified and hence relatively rare events can be identified almost as easily as more common events. This 'biological' purification technique sets DNA apart from all other types of molecule which rely upon chemical purification methods. Thus, in many cases where synthetic DNA is to be cloned or involved in a cloning event the DNA need not be 100% pure for its effective use. Moreover, many such applications require only tiny quantities (<1 μg or 0.1 nmol) of synthetic oligodeoxyribonucleotides. In contrast, much larger quantities of highly purified DNA are needed for structural studies involving, for example, n.m.r. spectroscopy or X-ray crystallography.

2.1 Construction of Duplex DNA

It is now possible to construct DNA duplexes of lengths up to several hundred

Introduction to DNA Synthesis

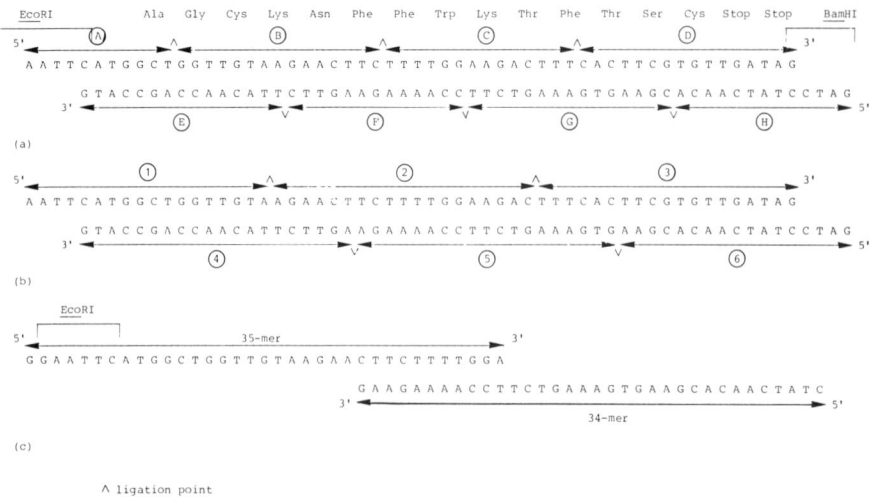

Figure 1. Alternative strategies for construction of duplex DNA.

base pairs. Applications vary from short linker sequences to large genes. Much use is made of the enzyme T4 DNA ligase to join together short, chemically synthesised oligonucleotides to form duplexes, which are then joined to biologically derived DNA. Particularly efficient joining takes place by sticky-end ligation. Two ends of DNA containing single-stranded overhangs, which are exactly complementary to one another, can be annealed and joined enzymatically to form a perfect duplex. At higher enzyme and DNA concentrations fully base-paired ends of DNA can be joined (blunt-end ligation).

In the first synthesis of a gene for a peptide hormone (somatostatin) carried out a few years ago (3) the chosen design (*Figure 1a*) made use of a strategy, based on the original gene synthesis approach of Khorana, involving the ligation of eight overlapping chains of 11−15 residues in length (A−H). Each chain was first phosphorylated at its 5′ end using T4 polynucleotide kinase and ATP and the ligations carried out in two stages. Chains A, B, E and F were joined in one experiment and C, D, G and H in another. In a second step the two duplexes were mixed and ligated. Finally the completed duplex was joined to the vector pBR322 DNA using the *Eco*RI and *Bam*HI sticky ends. The use of these two different restriction endonuclease sites allowed a unique orientation of the synthetic fragment to be obtained. Only ∼0.2 μg of duplex was required for joining to pBR322 DNA in a 50 μl reaction volume (∼0.1 μM), whereas blunt-ended ligation would have required at least 1 μM of ends for reasonable efficiency. Thus, for duplex construction by the overlap method a few micrograms of each oligonucleotide is sufficient.

Now that longer oligonucleotides are readily accessibly by chemical synthesis, a more modern approach might be that shown in *Figure 1b*. Overlaps are still 5−6 bp but ligation of all six chains would probably be attempted in one step. The 5′ end fragments (1 and 6) are usually not phosphorylated until after the ligation step to circumvent dimerisation. Because 5' recessed ends do not phosphorylate well, duplexes are always arranged to avoid this configuration.

A radically different procedure of duplex DNA synthesis has been described (4). Two very long oligonucleotides are annealed at their 5' ends and the enzyme DNA polymerase I (Klenow subfragment) or AMV polymerase (reverse transcriptase) is used to fill in single-stranded sections, each chain acting as a primer for the other strand. *Figure 1c* shows how this might be applied to the somatostatin gene. The sticky ends would be obtained by blunt-end ligation with a *Bam*HI linker (see below) and digestion with both *Eco*RI and *Bam*HI endonucleases. Once again only microgram quantities of oligonucleotides are needed for this method. The route shown in *Figure 1c* requires only two thirds the amount of chemical synthesis of that shown in *Figure 1b* and at first sight looks highly attractive. However, the efficiency of the primer extension step is unpredictable, perhaps because of unwanted secondary structures in long single strands. Also, yields of chemically synthesised 34- or 35-residue oligonucleotide chains are relatively low and purification at this length relies primarily on size separation on polyacrylamide gels (Chapter 6). At this chain length it is more difficult to assay for the presence of modified nucleotide bases obtained as side reactions in chemical synthesis. Such modifications might lead to incorporation of a wrong nucleotide by route *1c*. Route *1b* tends to eliminate the majority of base-modified oligonucleotides, owing to reduced thermal stability of mismatched duplexes. There is no published evidence on the comparative efficiency of the two routes and the proportion of clones containing genes of incorrect sequence in each case. However, as synthetic methods for longer chains improve, route *1c* becomes a reasonable choice.

Procedures for duplex DNA synthesis by fragment ligation are not covered in this book but have been well documented elsewhere (5).

For 5' phosphorylation see Chapter 6 and for a discussion of ways of avoiding incorrect duplex formation see Section 3.1

Blunt-end ligation is commonly used for introduction of a restriction endonuclease site on the end of a piece of DNA. A short self-complementary oligonucleotide (linker) (*Figure 2*) is ligated, after 5' phosphorylation, to blunt-ended DNA using high concentrations of enzyme and linker (>1 μM) DNA. A mixture of products with varying numbers of linkers on each end is obtained but, after cleavage by restriction endonuclease, the DNA is converted to sticky-end form. In contrast to longer duplexes, linkers are readily obtainable at less than 1 μM concentration and most experiments need less than 1 μg of linker. A range of linkers and other types of short adapter molecules (6) is widely available commercially. Protocols for blunt-end ligation have been published (7).

2.2 Oligodeoxyribonucleotides as Primers

Oligonucleotides will hybridise with complementary sequences on DNA and RNA templates and act as primers for synthesis of a complementary DNA strand. This has been exploited in a number of ways (8). Of widespread use is the direct sequencing of DNA and RNA by the 'dideoxy' chain termination method (9)

Introduction to DNA Synthesis

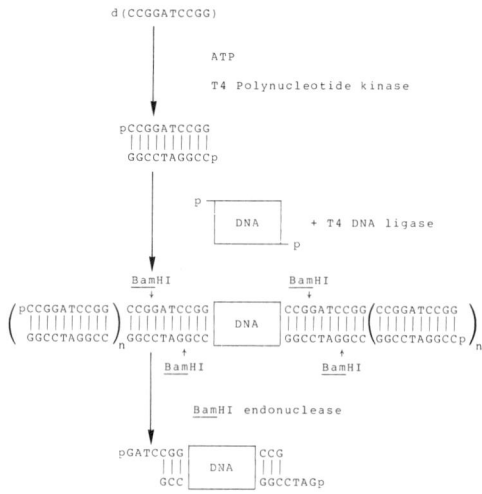

Figure 2. The addition of *Bam*HI linkers to DNA by blunt-end ligation.

using relatively pure DNA and RNA templates. The sequence is read off by positional relationships of radioactive bands representing chain terminated extension products, which are separated on polyacrylamide gels. There is no cloning step after the use of the oligonucleotide and hence there is an absolute requirement for specific priming. Careful choice of priming site to avoid priming on closely related sequences, limiting the molar excess of primer over template and high purity of oligonucleotide, are pre-requisites for reproducible results. With single-stranded templates, priming ability improves as a function of oligonucleotide length, but ensuring high purity becomes more difficult. Current usage of chains 15–20 residues long probably represents a compromise, although primers as short as 7-residue oligonucleotides have been used successfully.

Where the required DNA or RNA template represents only a small fraction of the total DNA or RNA in a mixture, an oligonucleotide primer can be used to enrich for the desired sequence. In the case of low abundance messenger RNA (mRNA) the reverse-transcribed product (cDNA) is either cloned directly or used as a hybridisation probe (see Section 2.3) to select its complement from a cDNA library generated by another method. To maximise the selectivity of a priming event a unique sequence primer is required. In the absence of any RNA sequence information, knowledge of part of the sequence of an encoded protein can be used to predict suitable primers. Selection is based either on statistical evidence of codon usage or on deliberate choice of G rather than A in the third position of codons for Glu, Gln and Lys, and C rather than U for Phe, Tyr, Cys, His, Asp and Asn, in the prediction that dT:rG and dG:rU base pairs would have only a marginal destabilising effect on DNA:RNA duplexes.

There are few known examples, however, where mismatched primers were successful in priming on the correct mRNA. More commonly a successful primer has been found later to have been exactly matched either by fortuitous choice or by synthesising and independently testing all possible primers corresponding to a particular sequence. An 11-residue oligonucleotide is thought to be the minimum length for observing a priming event. Often there are too many possible primers for the method to be practicable even though only tiny quantities (<1 μg) of each sequence are required. The alternative mixed-sequence probe approach (Section 2.3) is now generally preferred.

2.3 Selection of Recombinant DNA

The use of oligonucleotides as hybridisation probes (8) has for the most part supplanted primer extension methods for the cloning of specific DNA fragments. Collections of cDNA or genomic DNA sequences are prepared as 'shotgun libraries' by enzymatic or mechanical fragmentation of the DNA. This is followed by cloning of the product mixture into plasmids, bacteriophages or hybrid vectors (e.g., cosmids, phasmids) *via* transformation or infection of *Escherichia coli* cells. Many thousands of clones spread out on agar plates are screened by immobilisation on nitrocellulose filters and hybridisation of the DNA with radioactively labelled probes. In rare cases synthetic probes can be uniquely designed using neighbouring sequence information or by characterisation of frame-shift mutants of encoded proteins. More commonly, protein sequence information is used to determine all possible DNA sequences of a particular gene and a synthetic mixture of sequences (often 8, 16 or 32 chains) is used as the probe under stringent hybridisation conditions. Only the correctly matched sequence remains annealed to the cloned DNA when the hybridisation temperature is varied close to the melting temperature (see Section 3.1) thus identifying the clone containing the desired gene. *Figure 3* shows a particularly unusual case where the combined use of two 17-residue mixtures of 128 oligonucleotides each was used successfully to select clones containing cDNA corresponding to human apolipoprotein E (10). For the synthesis of mixed sequence probes by the solid-phase method (see Section 4.2) mixtures of two or four protected deoxyribonucleotides can be added simultaneously in a coupling reaction. Although one report suggests that adenine and guanine derivatives react slightly more slowly than cytosine and thymine derivatives (11), fortunately, the selection of clones is not critically dependent on precisely equal representation of all sequences. The characterisation of products from a mixed sequence synthesis is more complex. Polyacrylamide gel electrophoresis (Chapter 6) offers little or no resolution of an equal length mixture and hence is often useful for product isolation. H.p.l.c. (Chapter 5) may partially resolve simple mixtures and can sometimes be used to assess ratios of particular products. Maxam-Gilbert and wandering spot sequencing (Chapter 6) can be helpful in detecting if a particular sequence is absent. However, no current procedure can identify and quantify a mixture of 128 oligonucleotides. Fortunately, in this application of synthetic DNA the obtaining of a 'clean' product is perhaps of less consequence than in any other, and length assessment is usually sufficient.

A novel alternative for screening genomic libraries is the use of very long synthetic DNA probes. Here a unique sequence, deduced from its corresponding protein sequence by statistical analysis of codon usage, is prepared synthetically

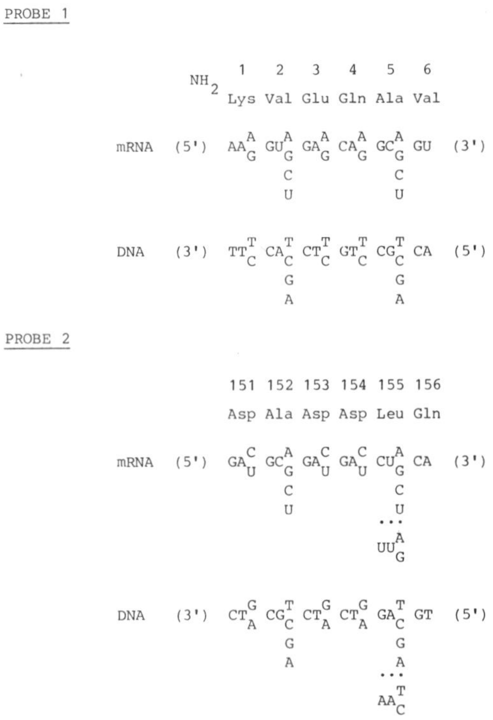

Introduction to DNA Synthesis

Figure 3. Mixed sequence probes designed for cloning the mRNA corresponding to apolipoprotein E.

as a duplex (Section 2.1). In a recent example, an 86-residue oligonucleotide probe, after cloning, successfully selected the gene for bovine pancreatic trypsin inhibitor from a genomic library. It proved later to be only 74% homologous to the natural sequence and there was no continuous homology of greater than 11 residues (12). Although it may soon be possible to make long synthetic probes by direct chemical synthesis, it is perhaps wiser to retain a cloning step both for purification and for verification of the DNA sequence.

2.4 Site-directed Mutagenesis

One method of obtaining a site-directed mutation in a DNA sequence is to construct a duplex (see Section 2.1) containing the mutation in both DNA strands and to insert the duplex into a gene whose corresponding wild-type sequence has been deleted, for example, by restriction endonuclease digestion. Such specific deletions and insertions are not always easy to obtain. A more general mutagenesis technique makes use of a mismatched primer hybridised to the wild-type DNA, obtained as a clone in single-stranded M13 viral DNA, at a temperature below the melting temperature of the mismatched duplex. For transitions and transversions a primer containing a single base-pair mismatch (*Figure 4a*) can be made reasonably specific. This is done by relative assessment of the stabilities on the other sites on the DNA, by keeping primer length and base composition

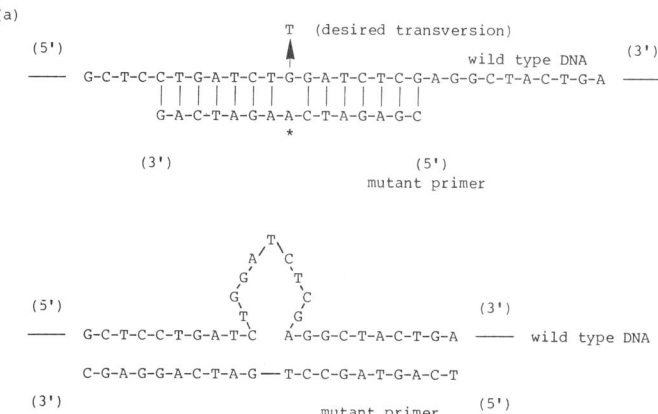

Figure 4. Hypothetical examples of oligonucleotides used to make (**a**) a transversion mutant, and (**b**) a deletion mutant.

within relatively narrow limits, and by keeping the mutant base centrally placed in the oligonucleotide (13). Primer extension using DNA polymerase is followed by a cloning step. Clones containing the mutant DNA, derived by segregation of the primer-extended strand, are selected. Selection is by phenotype or, where no phenotype is associated with the mutation, by use of the mutant primer as a hybridisation probe under more stringent conditions (Section 2.3). Double and multiple point mutations can also be constructed by use of mismatched oligonucleotides (14).

Mutant primers are generally 12–19 nucleotides in length and are required only in minute quantities (<1 µg). Since a cloning step is involved, high purity of mutant oligonucleotide is probably not essential but may help to ensure a higher frequency of desired mutation. The same general procedure has been useful in constructing deletions (*Figure 4b*). Here the oligonucleotide is designed to pair with sequences on both sides of the sequence to be deleted and again acts as a primer for synthesis of a second DNA strand. 20-Residue oligonucleotides have been used to delete short sequences (14–22 bases) (15). Another report claims 2% efficiency in deletion of 1143 bases using a synthetic 18-residue oligonucleotide (16). In general, however, such deletions are best made with longer oligonucleotides (20–30-residue oligonucleotides) in order to obtain a reasonable mutation frequency.

2.5 Structural Studies Involving Synthetic DNA

X-ray analysis of single crystals of short DNA duplexes derived from self-complementary synthetic oligonucleotides has produced some surprises. These include the discovery of the 'Z' DNA conformation obtained with the hexamer d(C-G-C-G-C-G) (17) and remarkable details of the 'B' conformation from the dodecamer d(C-G-C-G-A-T-A-T-C-G-C-G) (18). Reliable parameters for producing good crystals that will diffract to high resolution remain a mystery, and few sequences have been amenable to X-ray analysis. Sequence-specific and end

Introduction to DNA Synthesis

effects presumably predominate but high purity of the oligonucleotide is essential. Initial crystallisation attempts require only a few milligrams of material but on average about 30 mg is needed to grow a usable crystal even in favourable conditions. Similar problems confront those attempting co-crystallisation of DNA and protein. A co-crystal of the bacteriophage 434 repressor protein with a synthetic 14-base duplex DNA is one of the first to produce exciting results in this area (19).

Recent improvements in n.m.r. spectroscopy, particularly two dimensional proton n.m.r., has opened up a new dimension for studying DNA-protein recognition. Assignment of proton resonances has already been made for a non-symmetrical 17 residue oligonucleotide duplex (20). It will not be long before specific tertiary interactions between protein and DNA are studied by n.m.r. techniques. At least 20 mg of DNA duplex is required for a single n.m.r. experiment, but very high purity is less important than for crystallisation, because small impurities are not usually noticeable in an n.m.r. spectrum.

For preparation of oligonucleotides up to about 50 mg, scale-up of a solid-phase synthetic route is practicable and reasonably economical (Chapters 3 and 4) but not yet routine. Above 50 mg a classical solution synthesis is more appropriate but relatively labour intensive. In both cases, isolation of the oligonucleotide is time consuming.

3. DNA STRUCTURE AND CHEMICAL REACTIVITY

Before attempting to synthesise DNA it is useful to consider its structure and chemical reactivity and the consequences for possible chemical synthesis strategies. From a practical point of view the behaviour of DNA can be summarised in a few important principles, although some readers may prefer to consult more comprehensive texts (21,22).

3.1 **DNA Primary and Secondary Structure**

The primary structure of DNA is well known (*Figure 5*). It consists of a chain of 2-deoxy-D-ribose rings linked by $3' \rightarrow 5'$ phosphodiesters, carrying one of four possible heterocyclic bases in a β-configuration at each $1'$ sugar position. The high water solubility of DNA is due in part to the presence of ionisable phosphates and to the exocyclic primary amino groups in cytosine, adenine and guanine, which have strong hydrogen-bonding potential to solvent and to other polar molecules. For the same reasons DNA is relatively insoluble in organic and non-polar solvents. Hence, purification of DNA must be carried out in primarily aqueous solution where its polyionic character can be exploited by use of ion-exchange chromatography or electrophoresis. For chemical synthesis of DNA, however, it is necessary to use totally non-aqueous conditions. To achieve this the DNA must be prepared in modified form, where both phosphates and primary amino groups are chemically protected, and then the protecting groups removed selectively to liberate the unmodified DNA.

Another consequence of the hydrogen-bonding potential of the primary amino groups is the formation of secondary structures. In particular, these involve

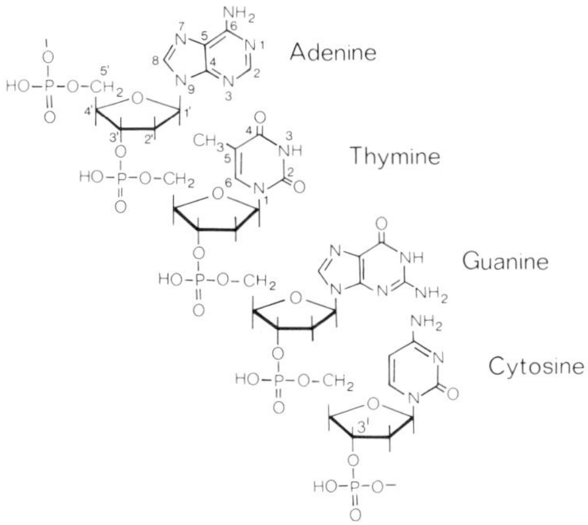

Figure 5. The primary structure of DNA showing the four common heterocyclic bases and the numbering system.

specific base pairing of guanine with cytosine and adenine with thymine, which can occur internally (to form hairpin loops) or intermolecularly (to form the double-stranded, natural form of DNA). Another stabilising force in such structures is base stacking, vertical base-base interactions, which are particularly strong in the case of purine-purine.

In chemical synthesis of oligodeoxyribonucleotides it is very important for such structure potential to be carefully considered before beginning any synthesis project. The main difficulties arise not in the synthesis itself, where the amino groups are chemically protected and are unable to initiate significant secondary structure, but during final purification of the deprotected oligonucleotide. Here, during chromatography, strongly denaturing conditions (e.g., formamide, urea) must be used, generally together with elevated temperatures. In extreme cases only polyacrylamide gel electrophoresis (Chapter 6) may adequately resolve the oligonucleotide. More importantly the desired application may be jeopardised. Unwanted secondary structures cannot usually be rectified, since the use of denaturing agents and elevated temperatures are precluded in most applications. Oligonucleotides with high purine content, in particular those containing long uninterrupted runs of guanine, require strongly denaturing conditions during purification (23). This is due to the tendency of such chains to aggregate in aqueous solution, especially at high ionic strength. The structure of such aggregates is unknown but their stability is presumably due to base stacking interactions rather than specific hydrogen bonds. Since there are no published studies on the effect of oligo(dG) aggregation on intermolecular hybridisation (specific inter-strand base pairing involved in many applications of synthetic DNA) long runs of guanine are probably best avoided.

A more common problem is the formation of unwanted base pairs. A useful exercise is the estimation of the T_m (the temperature at the mid-point of the transition between duplex and single-stranded forms; melting temperature) of desired interactions and comparison with those of other potential duplexes. For chains

Introduction to DNA Synthesis

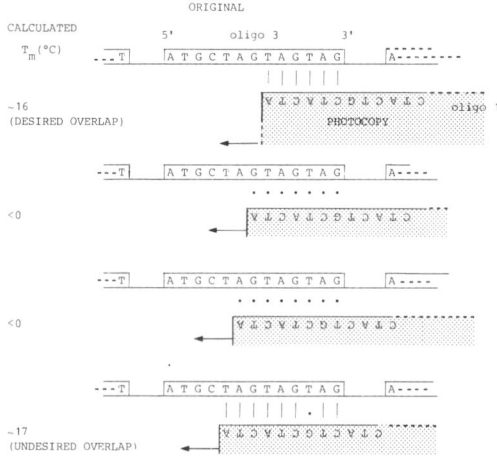

Figure 6. Visual inspection method for potential duplexes. Write down on the edge of a piece of graph paper the sequence of all chains in order (N.B. always write left to right for 5'→3'). Photocopy the page and align the edges of the two sheets so that the most 3' base on the original page is opposite (and inverted with respect to) the same 3' base on the photocopy. Slide the photocopy along so that the 3' base is now opposite the second base from the right on the original sheet. Continue sliding one base at a time inspecting for possible stable duplexes. With a little practice this can be done quite quickly. Particularly to be avoided are self-complementary ends in an oligonucleotide and duplexes with T_m close to that of the desired duplex. (dA:dT = 2°K; dG:dC = 4°K; dG:dT = −5°K; other mismatches or loops = −10°K.) For gene synthesis judicious codon choice or adjustment of the position of ligation points can usually eliminate such problems.

up to about 200 long, T_m^{-1} (° Kelvin) is approximately proportional to n^{-1}, were n is oligonucleotide length. From this relationship an empirical formula for T_m known as the 'The Wallace Rule' has been derived for perfect duplexes of about 15−20 units under high ionic strength (6 x SSC) of 2°K per dA:dT and 4°K per dG:dC. A DNA duplex can adjust conformation to allow for a mismatch or small loop-out but is destabilised by 5−10°K.

When oligonucleotides are to be used as hybridisation probes, a computer search for unwanted duplexes formed with vector DNA is usually best. It can be carried out using generally available software developed for matching information derived from DNA sequencing experiments. In the construction of short duplex DNA (Section 2.1) a visual inspection is often sufficient (*Figure 6*).

Predicting secondary structures of long, single-stranded oligonucleotides (>25) is more complex since intramolecular base pairing and intermolecular structures involving large or multiple loops-out may exist and are hard to spot. Routine use of such longer oligonucleotides in duplex DNA synthesis (Section 2.1) will depend on better predictive ability.

3.2 Chemical Reactivity of DNA

The phosphodiester oxyanions and the exocyclic primary amino groups need to be chemically protected. This is to obtain solubility of DNA in organic solvents (Section 3.1) and because they are the major nucleophilic centres, and would interfere with the process of forming an internucleotide linkage between a hydroxyl group of the deoxyribose in one nucleotide unit and the phosphate group in the other. Moreover, the heterocyclic bases (nucleobases) are prone to a wide range of chemical reactions. Even the mildest of reducing agents cause N-oxide formation of adenine and cytosine or even worse destruction. All the nitrogen and oxygen atoms of the heterocycles are potential sites for electrophilic attack (the N-7 positions of purines being particularly vulnerable). This rules out alkylation reactions (except those highly sterically hindered) but mild acylation is reasonably well tolerated and can thus be used to protect the exocyclic amino groups. Similarly, most of the non-bridged carbon atoms on the heterocycles are sites for hydrolytic attack, preventing the use of strong aqueous acids or alkali. Milder acidic conditions are also ruled out because the N-Cl' glycosidic bond in purine deoxynucleosides is labile (depurination), and even more so when adenine and guanine are acyl protected. Only extremely mild, usually non-aqueous, acids can be used with any reliability. In contrast, DNA is not vulnerable to mild basic conditions, the only exception being the slow deamination of cytosine to give uracil by hydroxide ion at high pH. Fortunately the phosphodiester linkage in DNA is fairly stable to both acidic and basic hydrolysis (cf. RNA). However, the protected form (phosphotriester) is not, and methods of specific deprotection without causing internucleotide cleavage are crucial to successful DNA synthesis.

Such considerations limit the range of useful chemical reactions for DNA synthesis to:

(i)   mild alkaline hydrolysis;
(ii) very mild acidic hydrolysis;
(iii) mild nucleophilic displacement reactions;
(iv) base-catalysed eliminations;
(v)   certain mild redox reactions (e.g., iodine or Ag^+ oxidations, Zn^{2+} reductive eliminations).

However, none of these can be regarded as totally safe. With so many nucleophilic and electrophilic centres on the heterocyclic bases, absolute specificity is virtually impossible. Reasonable selectivity can be achieved by careful control of reaction conditions, but the essential principle remains that a slight change in reaction conditions (e.g., solvent, polymer support, catalyst) can give a very large change in amount of side products. This should be borne in mind when comparing, for example, the use of a particular reagent in solution and in conjunction with different polymer supports.

4. CHEMICAL SYNTHESIS OF OLIGODEOXYRIBONUCLEOTIDES

The key step in DNA synthesis is the specific and sequential formation of internucleotide phosphate linkages. Since a deoxyribonucleoside monomer contains two hydroxyl groups (3' and 5'), one must be chemically protected while the other is specifically phosphorylated (or phosphitylated) and then coupled to the next deoxyribonucleoside unit. Meanwhile, other reactive moieties (e.g., exocyclic amino groups) must also be protected. From the earliest days strategies of

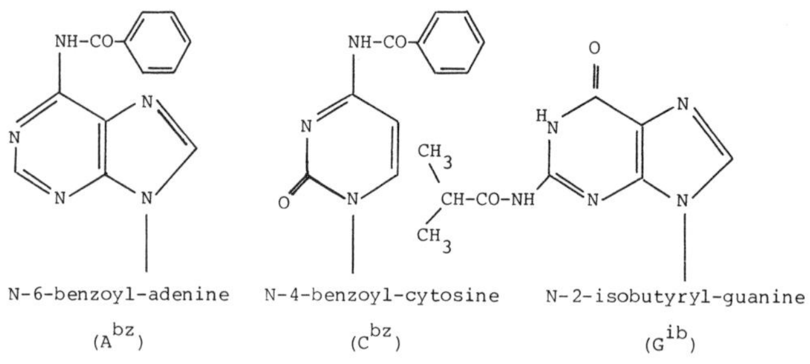

Introduction to DNA Synthesis

Figure 7. Exocyclic amino protecting groups for adenine, cytosine and guanine.

chemical synthesis have therefore revolved around the development of two kinds of protecting group:
(i) permanent, which remains attached to the oligonucleotide throughout the synthesis and is removed at the end of chain assembly;
(ii) temporary, which is only there to obtain specificity of a single reaction and is then removed immediately.

4.1 Protecting Groups

Heterocyclic bases require permanent protection at the exocyclic amino groups and this is achieved by acylation. Although several alternatives have been proposed, the most popular protecting groups remain benzoyl for adenine and cytosine and isobutyryl for guanine (*Figure 7*) (thymine usually requires no protecting group). These groups are stable to all the normal reactions used in oligonucleotide assembly but are cleaved at approximately equal rates by concentrated ammonia at the end of synthesis. Hitherto protection of the exocyclic amino groups (N-protection) was considered sufficient for the heterocyclic bases. More recently the characterisation of side reactions associated with guanine, especially significant during extended synthesis, has prompted the introduction of a second protecting group at the $O-6$ position (see Chapter 2).

Protecting groups at sugar hydroxyls can be temporary or permanent, depending on the synthesis strategy. For a number of reasons it is preferable for deoxyribonucleoside building blocks to be specifically phosphorylated (or phosphitylated) at the $3'$ position. Hence there is a need for temporary protection of the $5'$-hydroxyl group. This is currently fulfilled by one of a family of ether protecting groups (*Figure 8*) which are increasingly labile to acid in the order, trityl < monomethoxytrityl < dimethoxytrityl. Because of their bulk these protecting groups are introduced selectively at the $5'$ position. The dimethoxytrityl and a similarly useful pixyl group (see Chapter 4) are so extremely labile to acid (and needfully so to minimise depurination) that great care must be used in order to avoid inadvertent loss during chemical synthesis. Full details for the preparation of N- and $5'$-O-protected deoxyribonucleosides are given in Chapter 2.

For DNA synthesis by solution methods (24) (not covered in this book) a terminal $3'$-hydroxyl protecting group is necessary (cf. Chapter 7, where solution

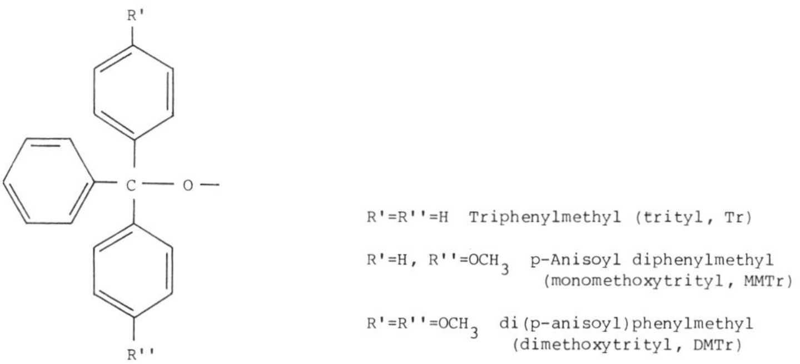

Figure 8. The 5'-O-triphenylmethyl group and its mono- and di-methoxy derivatives.

R'=R''=H   Triphenylmethyl (trityl, Tr)

R'=H, R''=OCH$_3$ p-Anisoyl diphenylmethyl (monomethoxytrityl, MMTr)

R'=R''=OCH$_3$ di(p-anisoyl)phenylmethyl (dimethoxytrityl, DMTr)

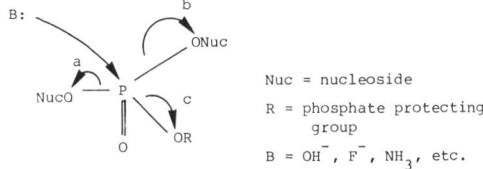

Nuc = nucleoside
R = phosphate protecting group
B = OH$^-$, F$^-$, NH$_3$, etc.

Figure 9. Undesirable mechanism for deprotection of phosphotriesters.

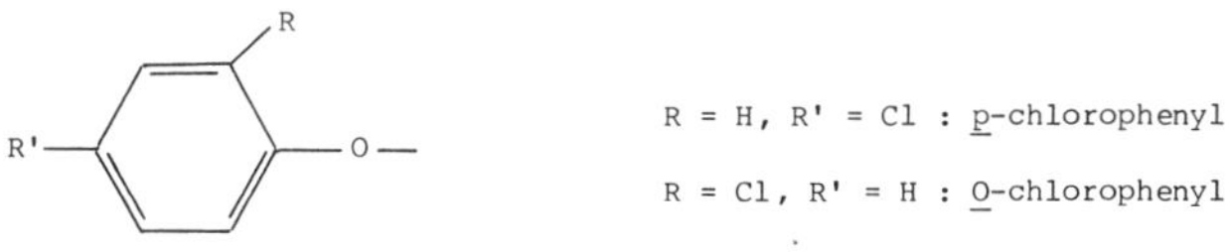

R = H, R' = Cl : p-chlorophenyl

R = Cl, R' = H : o-chlorophenyl

Figure 10. Chlorophenyl protecting groups for phosphate.

methods of RNA synthesis require a range of hydroxyl protecting groups). This is replaced in solid-phase synthesis by linkage to a solid support (a kind of permanent 'protecting group').

The choice of permanent protecting group for phosphate is dependent on the internucleotide coupling method. In the original work by Khorana no protecting group was used — the phosphodiester method (5). Chain extension was only possible because during coupling reactions phosphodiesters became temporarily blocked (to give substituted pyrophosphates) and this allowed eventual reaction of the sugar hydroxyl group. As might be expected, such reactions became progressively less selective for longer chains and the route is now only of historical and developmental interest. The development of permanent phosphate protecting groups selectively removable at the end of synthesis became of great importance. In particular, mechanisms of cleavage involving attack of a strong nucleophile (e.g., hydroxide ion) must be avoided (*Figure 9*), since they can lead to internucleotide cleavage (routes *a* and *b*) as well as the desired deprotection (route *c*). Useful phosphate protecting groups are the chlorophenyl derivatives (*Figure 10*) developed in association with phosphotriester chemistry and the methyl group

Introduction to DNA Synthesis

developed for use with phosphite-triester chemistry. The chlorophenyl group is cleaved selectively by use of a mild nucleophile, an oximate anion (Chapter 4) and the methyl group by the thiophenate anion (Chapter 3). In both cases the mechanisms of cleavage do not involve attack of a strong nucleophile at the phosphorus atom. The protecting groups are not interchangeable between chemistries since methyl-substituted phosphorylating agents are poorly reactive and it has been impossible to prepare chlorophenyl phosphitylating agents. However, newly proposed phosphitylating agents which contain permanent phosphate protecting groups cleavable by a β-elimination mechanism [β-cyanoethyl (25)] or by reductive elimination [1,1-dimethyl-β,β,β-trichloroethyl (26)] may supplant the methyl group in the future. The subject of specific 3'-phosphorylation and phosphitylation is too complex to be covered here but further discussion can be found in connection with coupling methods (Section 4.2.2).

4.2 The Solid-phase Method

The phosphotriester methods of DNA synthesis in solution (24) require coupling of one appropriately protected monomer unit containing a 3'-phosphate with another containing a 5'-hydroxyl group and separation of products and unreacted starting materials on a column of silica gel. Removal of a protecting group at one or other end gives a dimer block, which is coupled with another block to give a tetramer, and so on, to give larger protected fragments. Each coupling step requires a chromatographic purification and, although a skilled person can do this fairly quickly, adequate resolution of long chains is difficult and the method is labour intensive. It is therefore best applied to situations where very large quantities (>50 mg) of oligonucleotide are required.

Synthesis of oligonucleotides by the solid-phase method (27,28) brings advantages of speed, microscale operation, labour reduction and ease of mechanisation. In essence the method involves attachment of a 5'-O-protected deoxyribonucleoside to a solid support and chain assembly (*Figure 11*) by alternating terminal 5'-deprotection reactions (type 1) and coupling reactions (type 2). In both cases excess reagent is added to drive the reaction to completion and unreacted components removed merely by washing of the support with an appropriate solvent. Cycles of synthesis are continued until the required length is obtained and then the oligonucleotide is cleaved from the support, protecting groups removed, and the deprotected oligonucleotide purified. The four essential features of solid-phase synthesis are:

(i) support functionalisation;
(ii) chain assembly;
(iii) deprotection;
(iv) purification.

4.2.1 *Supports and Functionalisation*

Currently five types of material are applicable to solid-phase DNA synthesis (*Table 1*). All of these are suitable for the phosphotriester route but only four are

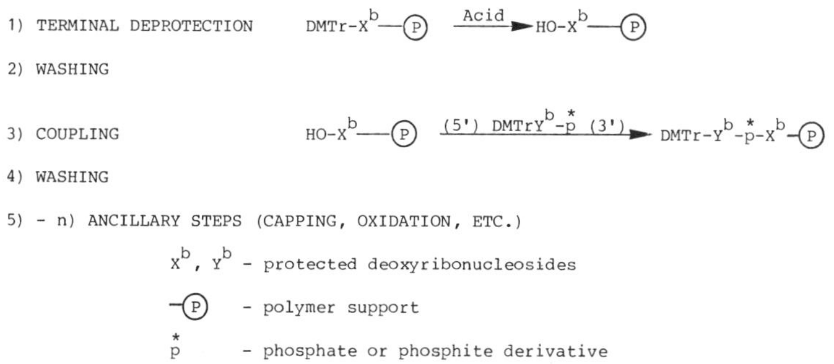

1) TERMINAL DEPROTECTION   DMTr-X^b—(P) $\xrightarrow{\text{Acid}}$ HO-X^b—(P)

2) WASHING

3) COUPLING   HO-X^b—(P) $\xrightarrow{(5') \text{ DMTrY}^b\text{-}\overset{*}{p}\text{ }(3')}$ DMTr-Y^b-$\overset{*}{p}$-X^b—(P)

4) WASHING

5) - n) ANCILLARY STEPS (CAPPING, OXIDATION, ETC.)

X^b, Y^b - protected deoxyribonucleosides

—(P) - polymer support

$\overset{*}{p}$ - phosphate or phosphite derivative

Figure 11. The essential terminal deprotection and coupling steps required in one cycle of chain assembly.

Table 1. Characteristics of Five Commonly Used Polymer Supports for Oligodeoxyribonucleotide Synthesis.

	Rigidity	Accessibility at low loading <0.1 mmol/g	Accessibility at high loading >0.1 mmol/g	Use for phospho-triester	Use for phosphite-triester
Polystyrene (1% DVB)	+	+ +	+	+ +	+
Silica gel	+ +	+ +	+/−	+	+ +
Glass beads (CPG)	+ +	+ +	+/−	+ +	+ +
Polyamide/Kieselguhr	+	+ +	+	+ +	−
Cellulose paper	+ +	+	−	+	+

+ + excellent; + good; +/− moderate; − poor.

useful in phosphite chemistry. Silica gel and glass beads are particularly useful on a microscale in both manual and mechanised operation because of their low loading (usually 30 − 70 μmol/g), high mechanical strength and lack of ability to swell. Polystyrene and kieselguhr/polyamide (a composite of polydimethylacrylamide gel embedded in macroporous kieselguhr) have advantages on a larger scale because, while reasonably rigid, their ability to swell allows better accessibility of sites when more highly loaded with oligonucleotide (>100 μmol/g). Cellulose paper has a special application for simultaneous synthesis of a large number of oligonucleotides (>20) on a microscale (Chapter 4). The dichotomy of a totally rigid support as opposed to one with highly accessible sites necessarily results in a compromise.

Despite a number of proposed alternatives only one type of linkage between support and deoxyribonucleoside has so far found universal utility. The 'succinate' linkage (*Figure 12*) is easily formed by coupling a deoxyribonucleoside 3'-O-succinate to an amino-substituted polymer. The linkage is rapidly cleaved by mild alkaline hydrolysis.

4.2.2 Chain Assembly

Perhaps most controversy and excitement has been generated by the two com-

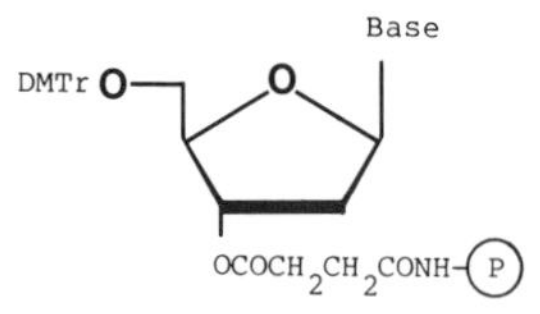

Introduction to DNA Synthesis

Figure 12. A 5′-O-dimethoxytrityl deoxyribonucleoside linked to a support *via* a 'succinate' linkage.

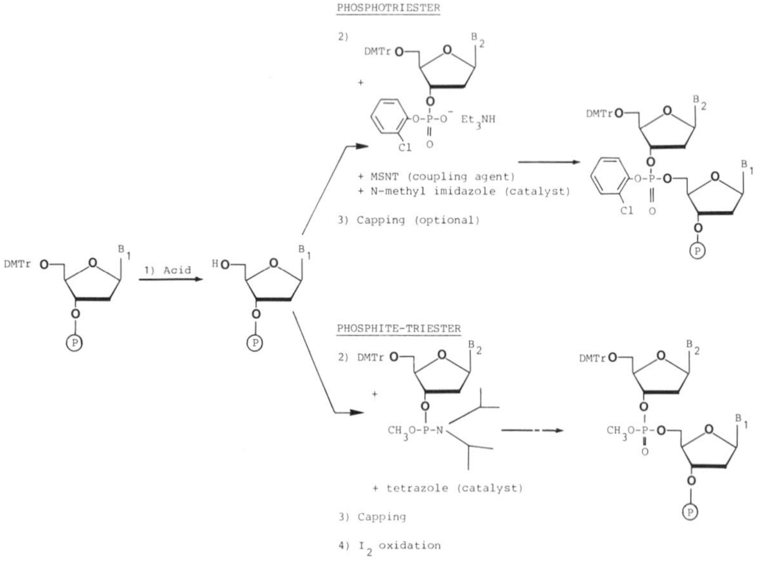

Figure 13. The phosphotriester and phosphite-triester chain assembly schemes.

peting chain assembly strategies, the phosphotriester and phosphite-triester (*Figure 13*). Yet dispassionate comparison shows more similarities than differences. The acidic terminal deprotection step (1) is identical in both routes. Weak protic acids (trichloroacetic, dichloroacetic) give deprotection rates independent of chain length but highly dependent on support and solvent. Those for Lewis acids (zinc bromide) depend considerably on chain length, support and solvent, but give rise to no undesirable depurination. The dimethoxytrityl cation released during deprotection is highly chromophoric and can be used as an indirect assay procedure. The remaining steps in the two assembly schemes yield phosphotriester products differing only in their permanent phosphate protecting groups even though the active mononucleotides added are quite different. In the phosphotriester route the 3′-O-phosphorylated monomer requires activation by a coupling agent and the coupling time is reduced to a few minutes by use of an extra catalyst. In the phosphite-triester route an active 3′-O-phosphitylated monomer (phosphorus in the more unstable P^{III} form) is prepared as a stable phosphoramidite but is converted to a highly active phosphitylating agent by an acidic catalyst. The intermediate dinucleoside phosphite is oxidised to the more stable phosphate (P^V) before chain extension. An extra capping step (blocking of

Table 2. Theoretical Overall Yield as a Function of the Number of Coupling Reactions and Average Repetitive Yield per Cycle.

Repetitive yield	Number of couplings					
	5	10	15	20	25	30
99	97.0	90.4	86.0	81.8	77.8	74.0
95	77.4	59.9	46.3	35.8	27.7	21.5
90	59.1	34.9	20.6	12.2	7.2	4.2
85	44.4	19.7	8.7	3.9	1.7	0.8
80	32.8	10.7	3.5	1.2	0.4	0.1

unreacted 5′-hydroxyl groups) is added in some methods as a precaution to aid final purification, although it is not strictly necessary to elaborate the desired oligonucleotide. The time required to complete each cycle does not differ greatly between the two routes.

4.2.3 Deprotection and Purification

It is essential for the protecting groups to be removed in the correct order. The phosphate protecting groups are cleaved first to form the corresponding phosphodiesters. Now ammonia can be safely used to cleave the N-acyl groups, and finally acid is used to remove the terminal DMTr group. The linkage between oligonucleotide and support is cleaved at the same time as chlorophenyl groups in the phosphotriester route but during ammonia treatment in the phosphite-triester method.

The use of a good purification scheme for the synthetic product is of vital importance to the success of the solid-phase method.

Several purification methods are given in Chapters 5 and 6. Some general points are worth mentioning, however. Firstly, because there are no chromatographic purifications during chain assembly, oligonucleotide impurities accumulate progressively as a function of chain length and all must be resolved from the desired product at the end of synthesis. Secondly, there is no single source of impurities. Truncated (prematurely terminated) chains, base-modified chains, and those resulting from chain cleavages during deprotection, all contribute. No single purification step can hope to resolve all of these, although sufficiently pure material for many applications in recombinant DNA can usually be obtained this way. Two independent methods of purification invariably give better results. Finally, the need to maintain high repetitive yields in synthesis is dramatically shown in *Table 2*. Although a 0.1% overall yield of a 30-residue oligonucleotide might be sufficient in terms of application requirement, identifying and purifying such a product from the unwanted 99.9% is complicated. Some tips on ways of maintaining good overall yields are given in the next section. Readers are warned that up to now yields in solid-phase syntheses have been reported in the literature in many different ways. To rectify this situation some new definitions are introduced in this book (*Table 3*). It is hoped these will be adopted in the future.

Introduction to DNA Synthesis

Table 3. Notes on the Definitions of Yields in Solid-phase Synthesis.

In this book the world 'yield' refers only to the amount of an isolated product. The word 'efficiency' is used for the amount of protected product while still attached to the support. Hence:

Assembly Efficiency (E_A): the amount of oligonucleotide in protected form in the product mixture while still attached to the support (estimated by indirect methods such as measurement of trityl group), compared with the original loading of first nucleoside on the support.

Coupling Efficiency (E_C): the assembly efficiency expressed on a per coupling (cycle) basis.

Overall Yield (Y_O): the amount of completely deprotected oligonucleotide obtained after isolation compared with the original loading of first nucleoside on the support. Yields of partially protected oligonucleotides are quoted without the word 'overall'. Yields are normally measured by direct methods (e.g., u.v. absorption of oligonucleotide calculated without correction for hypochromicity).

Repetitive Yield (Stepwise Yield or Coupling Yield) (Y_C): overall yield expressed on a per coupling (cycle) basis.

5. PITFALLS FOR THE UNWARY: PRACTICAL ADVICE TO THE NOVICE

Oligodeoxyribonucleotide synthesis is the first extended organic chemistry ever to be offered as a do-it-yourself method for non-specialists. Manual DNA synthesis procedures are now relatively straight-forward to follow and the use of machines is widespread. It should be recognised that a chemical synthesis method is distinctly different from a biochemical protocol and that a very slight change to a material or a method can often make the difference between barely obtaining a usable product and ensuring routinely reliable synthesis. This section focuses on aspects found by experience to be the most likely to cause trouble, particularly in chain assembly.

5.1 Solvents and Reagents

In an enzymatic reaction it is usual to run a control reaction in parallel to test the efficiency and specificity of an enzyme and its operating conditions. However, control DNA synthesis would be impracticable and time consuming. The synthesiser must therefore ensure the highest purity of batches of reagent and solvent. Rapid and fool-proof methods of analysis are not yet available for most materials used in chain assembly, but some simple checks can be helpful. In the author's experience the majority of synthesis failures are caused by impurities in reagents and solvents.

5.1.1 *Solvents*

Most commercially available solvents are not sufficiently pure for use in chemical synthesis. At least one purification step (usually a distillation) should be carried out. This is not a difficult procedure (Appendix I). Even batches of solvent supplied as 'specially purified for DNA synthesis' should be checked at least for colour (almost all liquids are colourless) and clarity (particulates will clog sinters, molecular sieve dust may cause synthesis failure – see below). Batch to batch variations in solvent quality are a major source of problems, and therefore redistillation of all solvents is recommended. The impurities of most concern are water, acids and bases, and metal ions.

Water. A single distillation from a water-absorbing chemical will reduce the water content of most solvents to 10 p.p.m. or less. This level is adequate for synthesis, since small quantities of water are allowed for in the coupling procedures. Many solvents are hygroscopic and therefore all solvent containers should be stoppered and opened only for brief periods. Storage under argon is not usually necessary.

Acids and bases. Chlorinated hydrocarbons often contain traces of hydrochloric acid which may lead to loss of dimethoxytrityl groups at inopportune moments during synthesis. This can be tested for by dissolving 1 mg of a 5'-O-dimethoxytrityl deoxyribonucleotide derivative in 5 ml of the solvent and measuring the u.v. absorbance at 495 nm. No absorbance should be obtained. Primary and secondary amines in polar solvents (pyridine, DMF, acetonitrile) will cleave the oligonucleotide-support linkage and can be assayed for by the ninhydrin test (Appendix I).

Metal ions. Although not a problem in the solvents, molecular sieve which is often added as a drying agent, contains large amounts of iron. Some solvents (e.g., pyridine) leach out iron (as ferric ion) from the molecular sieve which binds to DNA and, in an as yet uncharacterised manner, causes base modification and internucleotide cleavage. This is usually manifest only after protecting groups are removed at the end of a synthesis. Other metal ions (e.g., Zn^{2+}, Cu^{2+}) may be harmful.

5.1.2 *Reagents*

In chain assembly small quantities of highly reactive impurities in deoxyribonucleotide monomers are extremely troublesome since the use of a large excess of monomer amplifies the impurity effect. Assay is difficult but essential (see Chapters 3 and 4). Impurities in catalysts, coupling agents and deprotecting agents are generally of much less effect.

5.2 **Apparatus**

Chain assembly can be carried out using many different types of apparatus and the choice is largely a personal and financial one. All are in principle applicable to both phosphotriester and phosphite-triester chemistries. There are four categories.

Manual. The support is held in a sintered glass funnel connected to an aspirator or pump. Solvents and reagents are added manually by pipette or syringe (Chapter 3). A variation is the use of a syringe to entrap the support and solvents introduced *via* the syringe needle (29).

Semi-manual. The support is held in a small column or reaction chamber and solvents added and expelled *via* a manually operated valve system and pump or gas pressure device. Coupling mixtures are added manually (see Chapter 4).

Semi-mechanised. As for semi-manual except solvent addition is controlled by a timer device.

Introduction to DNA Synthesis

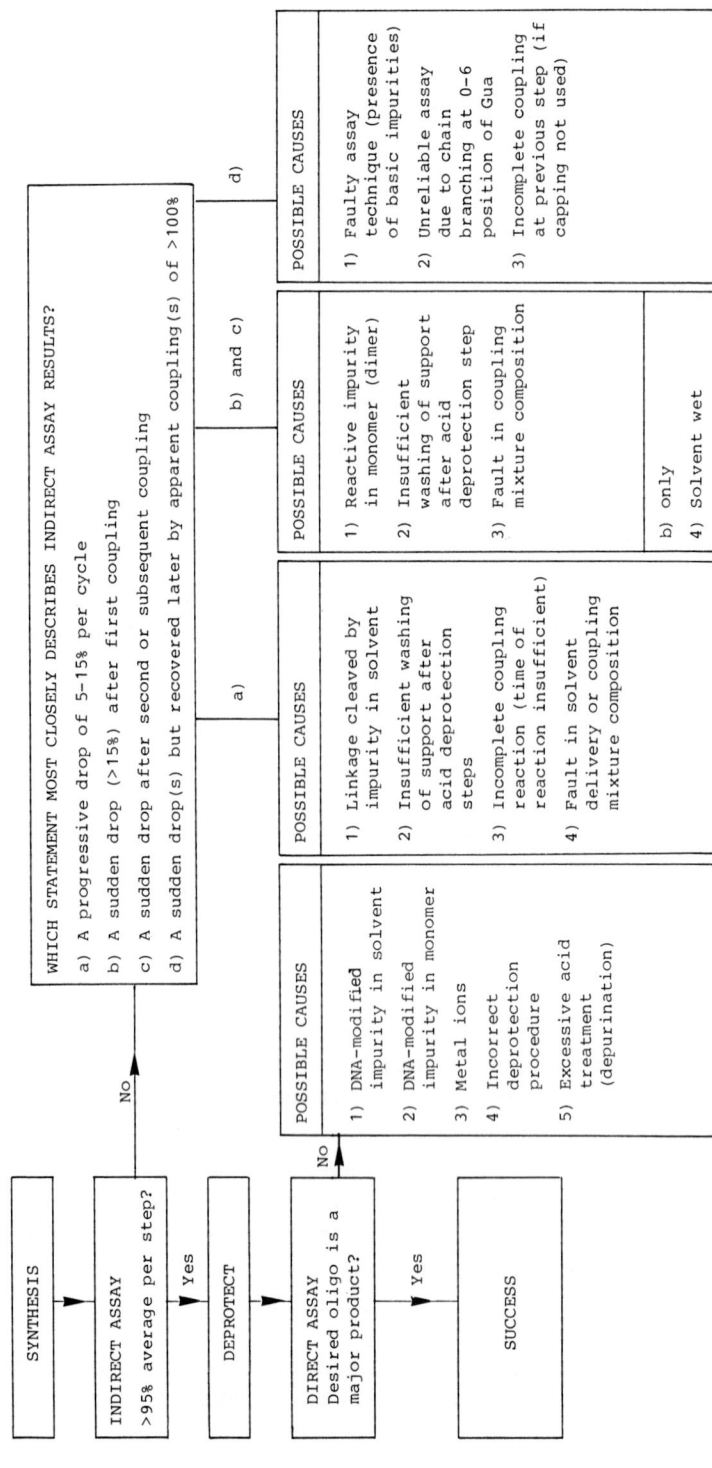

Figure 14. Flow sheet for trouble-shooting in solid-phase oligodeoxyribonucleotide synthesis.

Mechanised (DNA synthesis machines). All manipulations are controlled by a computerised program and manual intervention during synthesis is not required.

Comparison of manufacturers' equipment has been excluded from this book since specificiations will probably change considerably in the coming years. Some discussion of the various categories is given, however.

Machines do no more than one can do manually. They merely reduce the manual labour necessary for chain assembly. The time required per nucleotide addition is not substantially different for the four categories, but they vary in their operator control. Machines can make use of the time when an operator is absent (e.g., nights) and operate without getting tired. Some machines can handle three or four simultaneous syntheses, but manual procedures are better for a larger number of simultaneous syntheses.

Machines do not replace people. They change the nature of the work for an operator, removing some of the repetitive elements and replacing them with more work on oligonucleotide isolation and machine upkeep.

Machines make mistakes too. In manual synthesis human error is likely to occur. Machines are not infallible (especially when operated by humans), and they vary substantially in their capacity to check themselves. Even the most sophisticated does not yet have 'feedback control'. This is the capacity to analyse accurately the success of a synthesis and to act on the information. The machines, therefore, are not truly 'automated' (N.B. the measurement of release of dimethoxytrityl groups is merely a mechanical check and not a reliable guide to the amount of product obtainable, only to the efficiency of chain extension). A DNA synthesis machine should be thought of as a mechanical aid and not as an intelligent robot.

Finally, whichever category of apparatus is chosen, any part in contact with solvents should be constructed preferably of glass or Teflon. Even high quality stainless steel is slowly attacked by acid releasing ferric ion. Thus, if stainless steel is used, contact time with acids should be short.

5.3 Techniques and Trouble-shooting

Apart from the few exceptions clearly stated in Chapter 3, all techniques mentioned in this book can be handled by a non-specialist if care is given to the normal safety procedures used when dealing with chemicals (i.e., toxicity, flammability and corrosion). Personal protection (especially safety glasses) is recommended especially when using reduced pressure, bottled gases, pumping systems and distillations. When attempting a chemical technique for the first time, accuracy and care are more important than speed.

If the synthesis seems to have failed the trouble-shooting flow-sheet (*Figure 14*) may be of help. Finally, it is often quicker in solid-phase synthesis to re-synthesise the oligonucleotide (after determining the fault) than to attempt to identify and purify a product from a very poor synthesis.

6. REFERENCES

1. Khorana,H.G. (1968) *Proceedings of the Seventh International Congress of Biochemistry,* Tokyo, 1967, p. 17.

2. Davies,J.E. and Gassen,H.G. (1983) *Angew. Chem. Int. (Engl. Ed.)*, **22**, 13.
3. Itakura,K., Hirose,T., Crea,R., Riggs,A.D., Heyneker,H.L., Bolivar,F. and Boyer,H.W. (1977) *Science (Wash.)*, **198**, 1056.
4. Rossi,J.J., Kierzek,R., Huang,T., Walker,P.A. and Itakura,K. (1982) *J. Biol. Chem.*, **257**, 9226.
5. Brown,E.L., Belagaje,R., Ryan,M.J. and Khorana,H.G. (1979) *Methods Enzymol.*, **68**, 109.
6. Lathe,R.F., Lecocq,J.P. and Everett,R. (1983) in *Genetic Engineering 4*, Williamson,R. (ed.), Academic Press, p. 1.
7. Maniatis,T., Fritsch,E.F. and Sambrook,J. (1982) *Molecular Cloning. A Laboratory Manual*, published by Cold Spring Harbor Laboratory Press, N.Y.
8. Smith,M. (1983) in *Methods of RNA and DNA Sequencing*, Wasserman, S.M. (ed.), Praeger Scientific, NY, p. 23.
9. Bankier,A.T. and Barrell,B.G.(1983) in *Techniques in the Life Sciences B5*, **Vol. B508**, *Nucleic Acids Biochemistry*, published by Elsevier, Ireland, p. 1.
10. Wallis,S.C., Rogne,S., Gill,L., Markham,A., Edge,M., Woods,D., Williamson,R. and Humphries,S. (1983) *EMBO J.*, **2**, 2369.
11. Ike,Y., Ikuta,S., Sato,M., Huang,T. and Itakura,K. (1983) *Nucleic Acids Res.*, **11**, 477.
12. Anderson,S. and Kingston,I.B. (1983) *Proc. Natl. Acad. Sci. USA*, **80**, 6838.
13. Zoller,M.J. and Smith,M. (1983) *Methods Enzymol.*, **100**, 468.
14. Dalbie-Mcfarland,G., Cohen,L.W., Riggs,A.D., Morin,C., Itakura,K. and Richards,J.H. (1982) *Proc. Natl. Acad. Sci. USA*, **79**, 6409.
15. Chan,V.L. and Smith,M. (1984) *Nucleic Acids Res.*, **12**, 2407.
16. Osinga,K.A., Van der Bliek,A.M., Van der Horst,G., Groot Koerkamp,M.J.A., Tabak,H.F., Veenemar,G.H. and Van Boom,J.H. (1983) *Nucleic Acids Res.*, **11**, 8595.
17. Wang,A.H.-J., Quigley,G.J., Kolpak,F.J., Crawford,J.L., Van Boom,J.H., Van der Marel,G. and Rich,A. (1979) *Nature*, **282**, 680.
18. Drew,H.R., Wing,R.M., Takano,T., Broka,C., Tanaka,S., Itakura,K. and Dickerson,R.E. (1981) *Proc. Natl. Acad. Sci. USA*, **78**, 2179.
19. Anderson,J., Ptashne,M. and Harrison,S.C. (1984) *Proc. Natl. Acad. Sci. USA*, **81**, 1307.
20. Weiss,M.A., Patel,D.J., Sauer,R.T. and Karplus,M. (1984) *Proc. Natl. Acad. Sci. USA*, **81**, 130.
21. Tso,P.O.P., ed. (1974) *Basic Principles in Nucleic Acids Research*, Vols. I and II, published by Academic Press.
22. Kotchetkov,N.K., Budovskii,E.I. (eds.) (1972) *Organic Chemistry of Nucleic Acids, Parts A and B*, published by Plenum Press.
23. Newton,C.R., Greene,A.R., Heathcliffe,G.R., Atkinson,T.C., Holland,D., Markham,A.F. and Edge,M.D. (1983) *Anal. Biochem.*, **129**, 22.
24. Narang,S.A., Hsiung,H.M. and Brousseau,R. (1979) *Methods Enzymol.*, **68**, 90.
25. Sinha,N.D., Biernat,J. and Köster,H. (1983) *Tetrahedron Lett.*, **24**, 5843.
26. Letsinger,R.L., Groody,E.P., Lander,N. and Tanaka,T. (1984) *Tetrahedron*, **40**, 137.
27. Gait,M.J. (1980) in *Polymer-supported Reactions in Organic Synthesis*, Hodge,P. and Sherrington,D.C. (eds.), John Wiley, p. 435.
28. Wallace,R.B. and Itakura,K. (1983) *Chem. Anal.*, **66**, 631.
29. Letsinger,R.L. and Tanaka,T. (1982) *Nucleic Acids Res.*, **10**, 3249.

UPDATE

New techniques for the use of oligodeoxyribonucleotides as gene probes (section 2.3) include the incorporation of inosine as a 'wobble base' at ambiguous codon positions (30) and a procedure of hybridisation that allows stringency to be controlled as a function of chain length and irrespective of sequence (31).

The use of o-chlorphenyl as a phosphate protecting group in oligodeoxyribonucleotide synthesis using the phosphite-triester method (section 4.1) has now been reported (32).

30. Ohtsuka,E., Matsuki,S., Ikehara,M., Takahashi,Y. and Matsubara,K. (1985) *J. Biol. Chem.*, **260**, 2605.
31. Wood,W.I., Gitschier,J., Lasky,L.A. and Lawn,R.M. (1985) *Proc. Natl. Acad. Sci. USA*, **82**,1585.
32. Fourrey,J.L. and Varenne,J. (1984) *Tetrahedron Lett.*, **25**, 4511.

CHAPTER 2

Preparation of Protected Deoxyribonucleosides

ROGER A. JONES

1. INTRODUCTION

Chemical synthesis of ribo- and deoxyribo-oligonucleotides relies on precise manipulation of the reactive functional groups by the use of protecting groups. Nucleosides are multifunctional molecules with hydroxyl, amino and amide groups each of which may require protection, generally by different protecting groups. Some, such as the amino groups, must remain protected throughout the synthesis, while others, notably the 5'-hydroxyl group, must be deprotected after each elongation cycle. The number of different protecting groups that have been proposed is staggering (1–3). Yet for the most part the groups used, and the overall strategy employed, are those originally devised by Khorana and co-workers 20 years ago (4). In this approach the amino groups are protected by acylation; usually the benzoyl moiety (Bz) is employed for adenine and cytosine residues while the isobutyryl group (Ib) is used with guanine (*Figure 1*). These groups are stable to conditions employed during the course of a synthesis but are readily removed with aqueous ammonia at the end. The 5'-hydroxyl function is protected as an acid-labile trityl ether; either the 4,4'-dimethoxytrityl (DMTr) or the 4-methoxytrityl (MMTr) group is used. Because of the 10-fold greater rate of clevage of the DMTr *versus* the MMTr group the former is principally used in DNA synthesis, where depurination is a problem, while the latter is principally used in RNA synthesis, where the longer acidic treatment required does not pose any threat. The 3'-hydroxyl group (of a deoxyribonucleoside) is then available for phosphorylation or phosphitylation, depending on the method of synthesis to be employed, or for protection by reaction with benzoyl chloride or levulinic anhydride (1–3). For ribonucleosides a 2'-hydroxyl protecting group must also be used, most commonly the tert butyldimethylsilyl group (Chapter 7).

The main problems with this protection strategy in DNA synthesis are:

(i) depurination during cleavage of the DMTr group with protic acids;
(ii) degradation of guanine (and to a much lesser extent, thymine) during condensation reactions.

Depurination may be avoided by using zinc bromide ($ZnBr_2$) instead of protic acids, although $ZnBr_2$ reacts more slowly, and may therefore lead to incomplete detritylation (2,3). Guanine degradation results from the reactivity of the oxygen atom at the 6 position to a wide variety of reagents, including both condensing

23

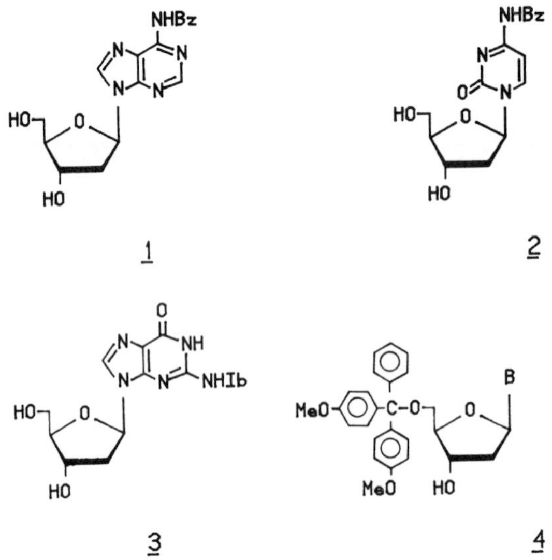

Preparation of Protected Deoxyribonucleosides

Figure 1. Protecting groups for exocyclic amino groups and 5' hydroxyl group. Bz = benzoyl; Ib = isobutyryl.

agents (5) and phosphorylating agents (6). This realisation has led to the development of a modified protection strategy for guanine that now includes O^6 protection (see Section 6) (7–10).

2. PER-ACYLATION

The preparation of protected deoxyribonucleosides is a matter of chemically differentiating between hydroxyl and amino groups. Only in the case of cytosine has it been possible to selectively N-acylate (1–3); the adenine and guanine amino groups are too weakly basic for selective reaction. However, it is possible to selectively de-acylate, that is, to differentiate by making use of the more rapid hydrolysis, at pH greater than 10, of the esters *versus* the amides. Thus, the classical procedure developed by Khorana and co-workers for both deoxyribo- and ribonucleosides, is to per-acylate the nucleoside, acylating both the hydroxyl groups and the amino group, and then to selectively hydrolyse the esters leaving the N-acylated nucleoside.

2.1 General Procedure for N-Acylation by Per-acylation

(i) Suspend the deoxyribonucleoside (10 mmol) in about 100 ml of dry pyridine (distilled from CaH_2, see Appendix I) contained in a 250 ml round-bottomed flask. Protect the reaction from moisture (drying tube) and cool it in an ice bath.

(ii) Add 10 equivalents of the acylating agent (either 11.6 ml of benzoyl chloride or 16.6 ml of isobutyric anhydride).

(iii) After 2 h pour the reaction mixture into 100 ml of cold, 5% aqueous sodium bicarbonate contained in a 1 litre Erlenmeyer flask (**caution**: there may be foaming).

(iv) Pour the mixture into a separating funnel and extract it with a 200 ml portion of ethyl actate. Drain out the lower (aqueous) layer and wash the ethyl acetate layer with a 50 ml portion of water. Drain out the wash and decant the ethyl acetate layer into a 1 litre round-bottomed flask. Repeat the extraction. Concentrate the combined ethyl acetate layers (rotary evaporator) to a gum.

(v) Dissolve the gum in 50 – 100 ml of a mixture of pyridine:methanol:water (65:30:5), cool in an ice bath and add an equal volume of cold 2 N sodium hydroxide solution of the same solvent composition.

(vi) After 20 min continue with either (a) or (b).

(a) Add excess pyridinium form sulphonic acid resin (Bio Rad AG50x2 H$^+$ form, or equivalent, washed with pyridine). Filter. Wash well. Extract with ether. Concentrate the aqueous layer until crystallisation begins.

(b) Add 1.2 equivalents of ammonium chloride (NH$_4$Cl) to neutralise (6.4 g NH$_4$Cl for 100 ml of 1 N NaOH). Concentrate the mixture until crystallisation begins.

The yields for this procedure are generally good (>70%). The most critical step is the controlled hydrolysis; a homogeneous solution is essential. The pyridine:methanol:water mixture is very reliable, but other mixtures, such as ethanol:water, can also be used. The use of NH$_4$Cl in step (vi)(b) eliminates the need to handle (and regenerate) the large quantities of ion-exchange resin required by (a). However, because of the ammonia liberated it is not possible to determine directly (by pH paper) that all of the hydroxide ion has been consumed. Use of a large excess of NH$_4$Cl should be avoided, as it may co-crystallise with the deoxyribonucleoside product.

3. TRANSIENT PROTECTION

The per-acylation method above works well, is reliable and gives reasonable yields. However, it is time consuming, because it requires isolation of the per-acylated intermediate, and the yield inevitably suffers because of the handling involved. An alternative, general procedure for N-acylation has been developed, based on 'transient protection' of the hydroxyl groups as trimethylsilyl ethers (11). Unlike ester groups, the trimethylsilyl ethers hydrolyse in aqueous pyridine (or with added dilute ammonia) without requiring a separate hydrolysis step (*Figure 2*; 5→7→8→1, and *Figure 3*; 10→2). Yields are generally high (85 – 95%).

3.1 General Procedure for N-Acylation by Transient Protection

(i) Suspend the deoxyribonucleoside (10 mmol) in dry pyridine (~100 ml) in a 500 ml round-bottomed flask. Protect the reaction from moisture (drying tube) and cool it in an ice bath.

(ii) Add 6.5 ml (~5 equivalents) of trimethylchlorosilane.

(iii) After 30 min add about 5 equivalents of acylating agent (either 6 ml of benzoyl chloride or 8.5 ml of isobutyric anhydride). Remove the reaction from the ice bath.

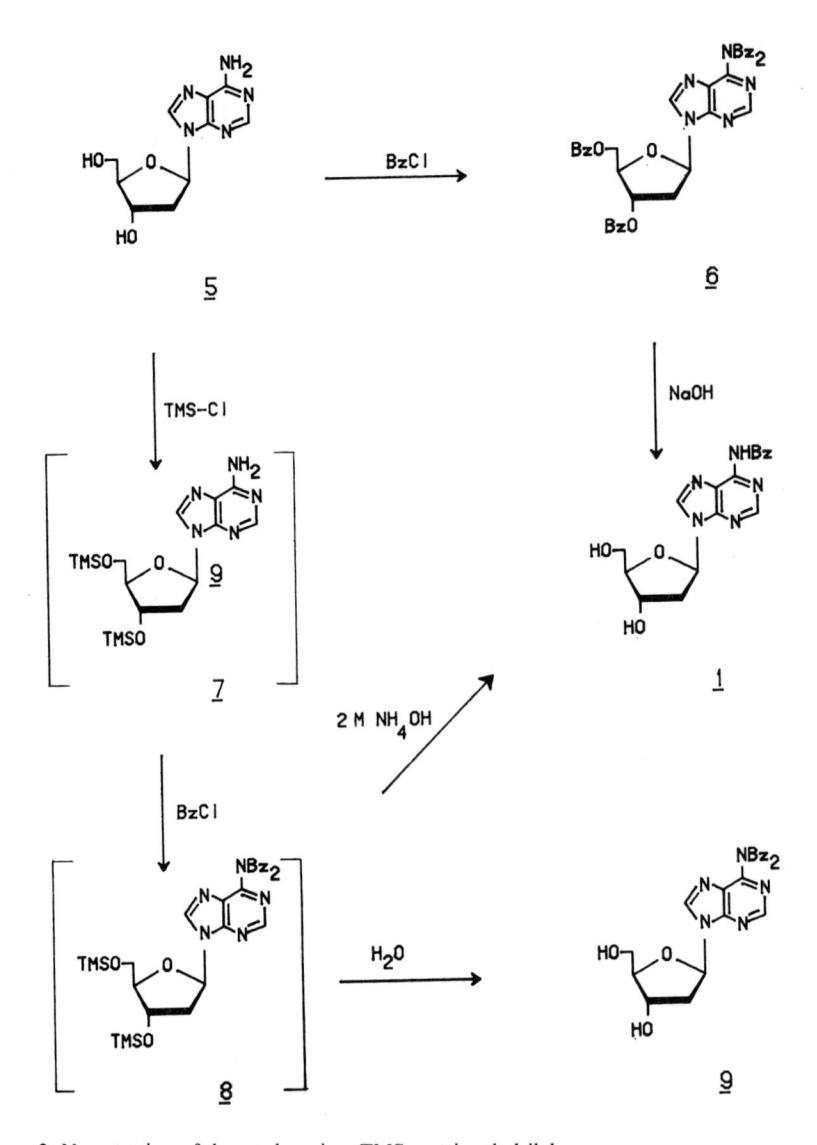

Preparation of Protected Deoxyribonucleosides

Figure 2. N protection of deoxyadenosine. TMS = trimethylsilyl.

(iv) After 2 h chill the mixture (ice bath) and add 20 ml of cold water followed after 15 min by 20 ml of concentrated aqueous ammonia (15 M) to give a solution approximately 2 M in ammonia.
(v) After 30 min concentrate the mixture to an oil (rotary evaporator).
(vi) Dissolve the oil in water (50–100 ml), wash the water with ether (separating funnel) and immediately separate the layers. Crystallisation should begin within minutes. Concentrate slightly (rotary evaporator) if crystallisation is not satisfactory (either no crystallisation or only a small amount).

Trimethylchlorosilane is quite inexpensive, but must be stored and handled carefully (use a chemically inert syringe) in order to obtain high yields, even though a considerable excess is used. Moreover, the trimethylchlorosilane should not be added all at once but over a few minutes. Too rapid addition can create sufficient localised heating, and acidity, for depurination to occur. This is particularly important for large-scale reactions and also applies to the addition of benzoyl chloride (either for N-acylation or for per-acylation). In a similar vein, quenching with water in step (iv) can cause significant warming, especially with large-scale reactions. Be sure that the reaction mixture is thoroughly chilled (ice bath) before quenching with cold water.

3.2 6-N,N-Dibenzoyl-2'-Deoxyadenosine

The reaction of adenine deoxyribonucleosides with benzoylating reagents almost always gives as the major product the 6-N,N-dibenzoyl derivative. The benzoyl chloride reaction used above in the per-acylation and in the transient protection approaches gives this product exclusively (*Figure 2*; 6, 8). There have been recent reports that these 6-N,N-dibenzoyl-2'-deoxyadenosine derivatives are more stable to the acidic conditions used for detritylation than are the monobenzoyl deoxyadenosine derivatives more commonly used (12). In any event, while both the per-acylation and transient protection methods initially give an N,N-dibenzoyl derivative (6 or 8), the 1 N NaOH treatment required in the per-acylation approach instantly cleaves one of these amino benzoyl groups. However, in the transient protection method, omission of the ammonolysis in step (iv) allows isolation of 6-N,N-dibenzoyl-2'-deoxyadenosine (9). Hydrolysis of the 3' and 5' trimethylsilyl groups is effected by the aqueous pyridine solution within a few hours.

4. PROTECTION OF THE 5'-HYDROXYL GROUP AS A 4,4'-DIMETHOXYTRITYL ETHER

4.1 **Reaction of N-Acyl Deoxyribonucleosides and Thymidine with 4,4'-Dimethoxytrityl Chloride**

The reaction of 4,4'-dimethoxytrityl chloride (DMTr-Cl) with an N-acyl deoxyribonucleoside or thymidine is highly selective for the primary, 5'-hydroxyl group over the secondary, 3'-hydroxyl (4). The reaction requires a slight excess of DMTr-Cl, and is usually complete within 1 h using 4-dimethylaminopyridine as catalyst (13). The product, isolated by solvent extraction, may be used without further purification, or may be readily purified by chromatography on silica gel (*Figure 3*; 2 → 12).

4.1.1 *General Procedure for Dimethoxytritylation*

(i) Dissolve the N-acyl deoxyribonucleoside (or thymidine) in dry pyridine (50 – 100 ml for 10 mmol of nucleoside). If the deoxyribonucleoside is not dry, a dry pyridine solution may be obtained by evaporating the pyridine (rotary evaporator; oil pump) and then adding fresh dry pyridine.

(ii) Add 0.05 equivalents (61 mg) of 4-dimethylaminopyridine (DMAP), 1.4 equivalents (1.9 ml) of triethylamine (TEA), and 1.2 equivalents (4.1 g) of 4,4'-dimethoxytrityl chloride (DMTr-Cl).

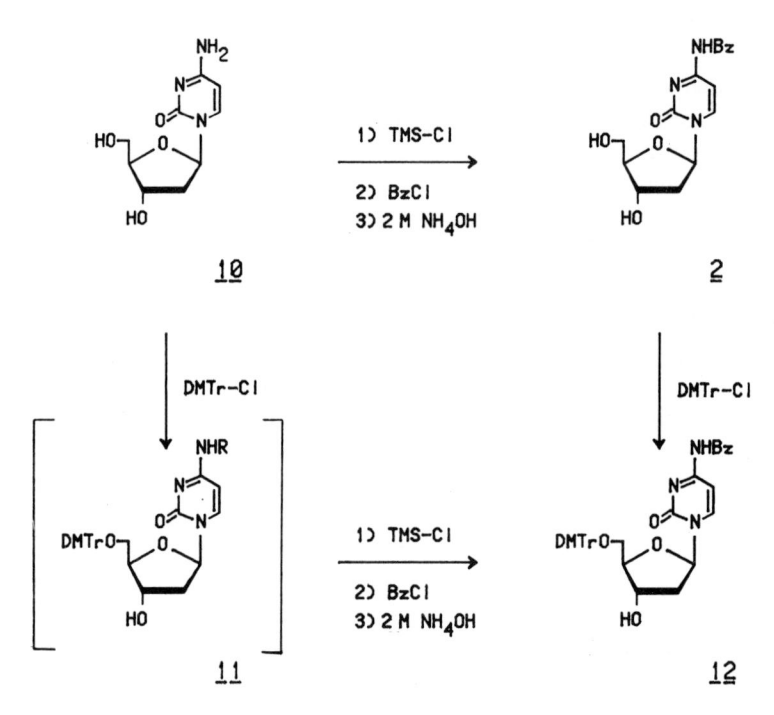

Preparation of Protected Deoxyribonucleosides

Figure 3. One-flask protection of deoxycytidine. DMTr = 4,4′-dimethoxytrityl; **(a)** R = H; **(b)** R = DMTr.

(iii) Follow the reaction by t.l.c. (silica; 2–10% methanol in methylene chloride). Add more DMTr-Cl and triethylamine if the reaction is not complete after 2 h.
(iv) When complete (2 h), add an equal volume of water and extract the product into two 500 ml portions of diethyl ether.
(v) Concentrate the ether layers to a solid foam (rotary evaporator, first water aspirator and then oil pump vacuum).
(vi) The DMTr-dT derivative can be crystallised from benzene (**warning**: carcinogen). The other DMTr derivatives may be purified by chromatography on silica gel. Use 100–500 ml of silica gel. The column is first eluted with methylene chloride and then with a step gradient up to about 5% methanol in methylene chloride (see Appendix I).
(vii) Combine the pure fractions and concentrate to a solid foam.

4.2 One-flask $O^{5'}$ and N Protection of Deoxyadenosine and Deoxycytidine

An alternative to the above procedure, which is effective with deoxyadenosine and deoxycytidine, is to tritylate the deoxynucleoside first, and to N-acylate second (*Figure 3*; 10→11→12) (11). The advantage is that these reactions can be carried out sequentially, without isolation of intermediates, in a one-flask procedure. However, chromatography is more frequently necessary than is the case with the procedure described in Section 4.1.1. Moreover, the yields are more variable, and generally lower, at scales over 10−20 mmol. Thus for large-scale work (>20 mmol), the two-step procedure may be preferable since chromatography can probably be avoided, while for moderate scales (<20 mmol) the overall speed and efficiency of the one-step approach may be more attractive. These 5'-DMTr deoxyribonucleosides are extremely sensitive; presumably the variable yields reflect problems with mass or heat transfer, or both, which may in turn lead to loss of dimethoxytrityl groups.

4.2.1 General Procedure for One-flask Protection

(i) Suspend the deoxyribonucleoside (10 mmol) in dry pyridine (~100 ml) in a 250 ml round-bottomed flask. Protect the reaction from moisture (drying tube).

(ii) Add 0.05 equivalents (61 mg) of DMAP, 1.4 equivalents (1.9 ml) of TEA, and 1.2 equivalents (4.1 g) of DMTr-Cl.

(iii) When the reaction is complete (t.l.c., silica, 2−10% methanol in methylene chloride), chill in an ice bath and add (slowly) 5 equivalents (6.4 ml) of trimethylchlorosilane, followed (after 15 min) by 5 equivalents (5.8 ml) of benzoyl chloride. Remove the reaction from the ice bath.

(iv) After 2 h (deoxyadenosine) or 8 h (deoxycytidine), chill the reaction mixture in an ice bath and add cold water (20 ml) followed after 5 min by 20 ml of concentrated aqueous ammonia to give a solution approximately 2 M in ammonia.

(v) Stir for 30 min, concentrate (rotary evaporator) to approximately 20 ml, add 500 ml of ether and wash the mixture with 100 ml of water.

(vi) Concentrate the ether layer and purify by chromatography as above.

Reaction of deoxycytidine with DMTr-Cl gives a mixture of the desired $O^{5'}$-DMTr deoxyribonucleoside (11a) and the $O^{5'}$ N^4-bis (DMTr) derivative (11b), but the subsequent long reaction with benzoyl chloride in step (iv) cleaves the N^4-DMTr group. No such N-trityl derivative has been found with deoxyadenosine. Deoxyguanosine, however, gives predominant N^2-tritylation, and reaction with isobutyryl chloride does not cleave the N^2-DMTr group. Thus this procedure is not applicable to deoxyguanosine. Yields are generally high.

5. O^6 PROTECTION OF DEOXYGUANOSINE

The presence of guanine residues in an oligonucleotide has long been correlated with low yields and obvious degradation (dark colours, fluorescent spots on t.l.c.) (1−3). The source of this degradation has been shown recently to lie in the reactivity of the oxygen atom at the 6 position toward a wide array of reagents, among them the condensing and phosphorylating agents commonly employed in oligonucleotide synthesis (5,6). Thus, it is clear that for the phosphate triester method and perhaps for the phosphite as well, some manner of protection of the O^6 amide oxygen is required. (This is also true of the uracil, and to a lesser degree the thymine, O^4 amide oxygens.) The protecting groups proposed to date include phosphinothioyl (7), 2-nitrophenyl (8), and a series of substituted ethyl groups (9,10). The most promising appear to be the cyanoethyl (CNE) and nitrophenyl-

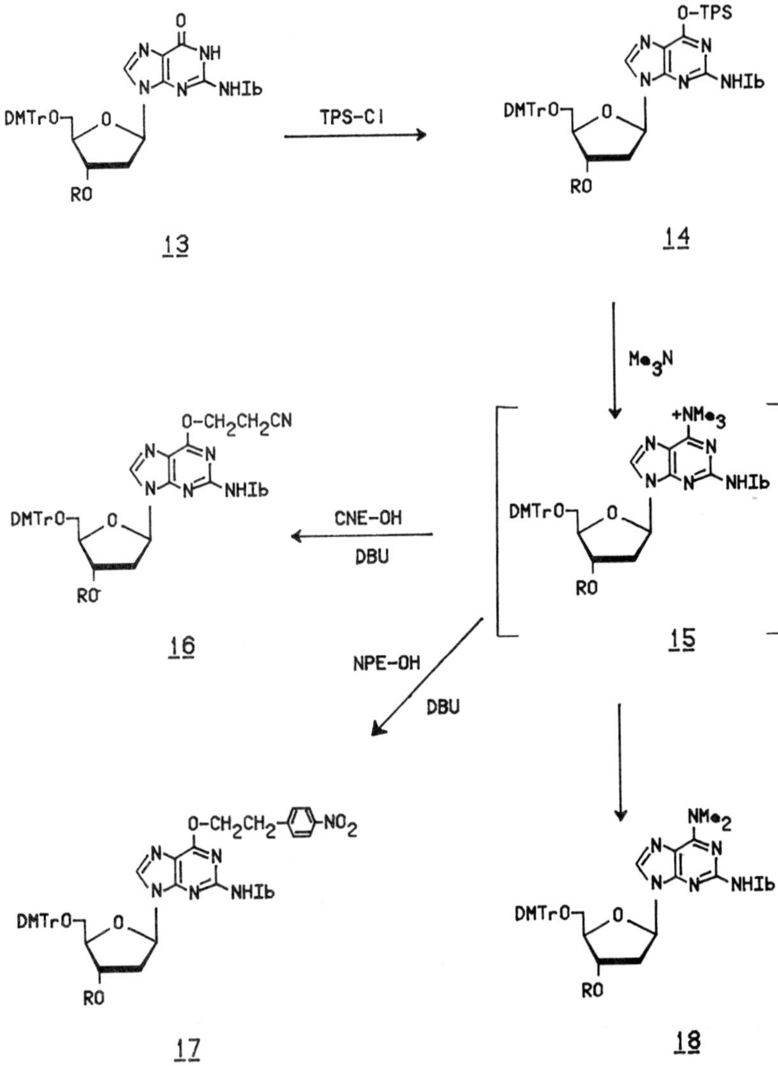

Preparation of Protected Deoxyribonucleosides

Figure 4. O⁶ protection of deoxyguanosine. TPS = triisopropylbenzenesulphonyl; **(a)** R = H; **(b)** R = COCH$_2$CH$_2$COCH$_3$; CNE-OH = cyanoethyl alcohol; NPE-OH = nitrophenylethyl alcohol; DBU = 1,8 diazabicyclo[5.4.0]undec-7-ene.

ethyl (NPE) groups. Each is cleaved by a β-elimination reaction: the CNE group is removed with either triethylamine or 1,8 diazabicyclo [5.4.0]undec-7-ene (DBU), while only DBU will cleave the more stable NPE group. Although the NPE group may be more stable than is required, it does have the advantage that both the 5'-DMTr derivative 17a (*Figure 4*) and O⁶-nitrophenylethyl deoxyguanosine (21) are crystalline compounds. Examples of the use of both 16 and 17 in oligonucleotide synthesis have been reported (14,15).

5.1 Synthesis of O⁶-Nitrophenylethyl or O⁶-Cyanoethyl Deoxyguanosine Derivatives by Sulphonylation/Displacement

Synthesis of O⁶ protected deoxyguanosine can be carried out as shown in *Figure 4*. Reaction of triisopropylbenzenesulphonyl chloride (TPS-Cl) with 13b gives the O⁶ sulphonylated derivative 14. Direct reaction of 14 with alcohols does not give 16, but only regenerates 13b. However, the TPS group is readily displaced with trimethylamine, to give 15, which in turn reacts with alcohols (in the presence of DBU) to give O⁶ alkylated derivatives 16 or 17.

5.1.1 *General Procedure for Sulphonylation/Displacement*

(i) Prepare an ethereal solution of levulinic anhydride by treating 60 mmol (6.1 ml) of levulinic acid with 30 mmol (6.2 g) of N,N'-dicyclohexylcarbodiimide (DCCI) in 30 ml of diethyl ether. Filter this solution into a pyridine solution (100 ml) of 0.05 equivalents (61 mg) of DMAP, and 10 mmol of 13a (prepared by the general procedures given above), in a 250 ml round-bottomed flask. Protect the reaction from moisture (drying tube).

(ii) After 1 h pour the reaction mixture into 200 ml of 5% sodium bicarbonate solution in a 1 litre Erlenmeyer flask. Extract the product into ethyl acetate, concentrate the ethyl acetate and purify the residue by chromatography on silica gel (methanol in methylene chloride eluant).

(iii) Dissolve 10 mmol of the above product (13b) in 50–100 ml of methylene chloride and add 0.05 equivalents (61 mg) of DMAP, 4 equivalents (5.6 ml) of triethylamine, and 2 equivalents (6.1 g) of TPS-Cl.

(iv) React for about 3 h. Add more TPS-Cl and TEA if the reaction is not complete (t.l.c., silica, ether:petroleum ether, 1:1). Then concentrate the reaction mixture to about one-half and either (a) add 4 equivalents (6.7 g) of 4-nitrophenylethanol and cool in an ice bath or (b) add 6 equivalents (4.1 ml) of 3-hydroxypropionitrile and cool to −20°C in an acetone/dry ice bath (use only small amounts of dry ice to maintain this temperature). Protect the reaction from moisture.

(v) Add 10–20 ml of trimethylamine, wait 10 min (ice bath) or 30 min (−20°C) and add 5 equivalents (7.5 ml) of DBU. After 30 min (ice bath) or 1 h (−20°C) add 20 ml of a mixture of acetic anhydride and pyridine (10% v/v).

(vi) After 10 min pour the reaction mixture into 100 ml of 5% sodium bicarbonate solution and extract the solution with three 250 ml portions of ether. Concentrate the ether layers then follow (a) or (b).
   (a) Purify the residue by chromatography for 16b/17b.
   (b) Treat the residue with 40 ml of 0.5 M hydrazine hydrate in a mixture of pyridine:acetic acid (4:1). Partition the mixture between water (200 ml) and ethyl acetate (200 ml), and wash the ethyl acetate layer with 50 ml of 5% sodium bicarbonate solution. Concentrate the ethyl acetate layer and purify the residue by chromatography (16a/17a). Crystallise 17a from ether.

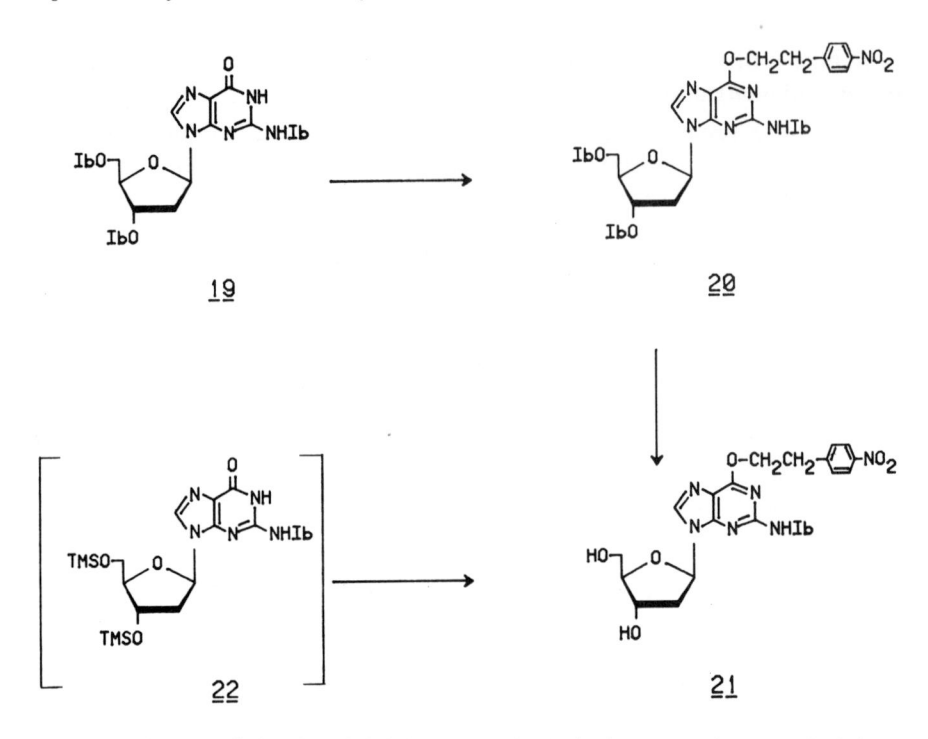

Preparation of Protected Deoxyribonucleosides

Figure 5. Synthesis of O⁶-nitrophenylethyl deoxyguanosine derivative 21 by Mitsunobu alkylation.

The amount of DBU employed (step v) is critical. Without sufficient DBU, displacement (15→16/17) will not occur; with too much DBU, deprotection (16/7→13) will occur especially with the cyanoethyl derivative 16. In addition, the trimethylammonium adduct 15 slowly demethylates to give the corresponding dimethylamino deoxyribonucleoside 18. Therefore it is important to effect the displacement with the alcohol as quickly as possible and to maintain a low reaction temperature. Acetylation of the excess alcohol in step (v) makes it much easier to remove during the subsequent chromatography. The overall yields are generally much better for the O⁶-nitrophenylethyl derivatives (60 – 70%) than for the O⁶-cyanoethyl derivatives (30 – 40%). In addition, 17a is a crystalline compound.

5.2 Synthesis of O⁶-Nitrophenylethyl Deoxyguanosine by Mitsunobu Alkylation

An alternative synthesis of the O⁶-nitrophenylethyl deoxyguanosine derivative 21 has been reported by Pfleiderer, based on Mitsunobu alkylation (*Figure 5*) (10,16). Treatment of triisobutyryl deoxyguanosine (19) with triphenylphosphine, diethylazodicarboxylate and nitrophenylethanol gives the O⁶ derivative 20. The 3′- and 5′-isobutyryl groups are then removed by a 3-day treatment with ammonia to give 21. Alternatively, 'transient protection' of the hydroxyl groups of N-isobutyryl deoxyguanosine (3) avoids the lengthy ammonia treatment required for ester hydrolysis, and gives crystalline 21 in 60 – 70% yield (*Figure 5*; 22→21) (17).

5.2.1 *General Procedure Using Transient Protection*

(i) Add 3.2 ml (22 equivalents) of trimethylsilyl imidazole (Aldrich) to a suspension of 10 mmol of N-isobutyryl-2'-deoxyguanosine in 50 ml of dioxane in a 250 ml round-bottomed flask. Protect the reaction from moisture.

(ii) After 30 min add 9.2 g (3.5 equivalents) of triphenylphosphine, 6.1 ml (3.5 equivalents) of diethylazodicarboxylate and 5.8 g (3.5 equivalents) of nitrophenylethanol.

(iii) After 1 h add 20 ml of 1 M pyridinium hydrofluoride in pyridine (prepared by adding 7.6 ml of 52% hydrofluoric acid to 50 ml of pyridine and concentrating to 20 ml), stir for 10 min and pour the reaction into 100 ml of 5% sodium bicarbonate solution.

(iv) Extract the mixture with methylene chloride, concentrate the extract and purify the residue by chromatography.

(v) Crystallise 21 from ether.

Tritylation of 21 to 17a can be carried out by the general procedures described above. It is also possible to use transient protection and Mitsunobu alkylation to prepare 17 from 5'-O-dimethoxytrityl-N-isobutyryl deoxyguanosine more directly. However in this case it is difficult to purify 17 from the triphenylphosphine oxide and diethyl azinodicarboxylate also produced in the reaction (17).

6. MOLECULAR WEIGHTS AND VOLUMES

2'-deoxyadenosine (dA) 251 g/mol
2'-deoxycytidine (dC) 227 g/mol
2'-deoxyguanosine (dG) 267 g/mol
2'-deoxythymidine (dT) 242 g/mol
benzoyl chloride (BzCl):116 ml/mol
isobutyric anhydride (Ib$_2$O):166 ml/mol
ammonium chloride (NH$_4$Cl):53.5 g/mol
trimethyl chlorosilane (TMS-Cl):127 ml/mol
4,4'-dimethoxytrityl chloride (DMTr-Cl):339 g/mol
4-dimethylaminopyridine (DMAP):122 g/mol
triethylamine (TEA):139 ml/mol
1,8 diazabicyclo[5.4.0]undec-7-ene (DBU):150 ml/mol
triisopropylbenzenesulphonyl chloride (TPS-Cl):303 g/mol
N,N'-dicyclohexylcarbodiimide (DCCI):206 g/mol
4-nitrophenylethanol (NPE-OH):167 g/mol
3-hydroxypropionitrile (CNE-OH):68 ml/mol
triphenyl phosphine (Ph$_3$P):262 g/mol
diethylazodicarboxylate (DEAD):174 ml/mol
levulinic acid:102 ml/mol
trimethylsilylimidazole:147 ml/mol

7. REFERENCES

1. Reese,C.B. (1978) *Tetrahedron,* **34**, 3143.
2. Ohtsuka,E., Ikehara,M. and Söll,D. (1982) *Nucleic Acids Res.,* **10**, 6553.
3. Narang,S.A. (1983) *Tetrahedron,* **39**, 3.
4. Schaller,H., Weiman,G., Lerch,B. and Khorana,H.G. (1963) *J. Am. Chem. Soc.,* **85**, 3821.
5. Reese,C.B. and Ubasawa,A. (1980) *Tetrahedron Lett.,* **21**, 2265.
6. Daskalov,H.P., Sekine,M. and Hata,T. (1980) *Tetrahedron Lett.,* **21**, 3899.
7. Daskalov,H.P., Sekine,M. and Hata,T. (1981) *Bull. Chem. Soc. Japan,* **54**, 3076.
8. Jones,S.E., Reese,C.B., Sibanda,S. and Ubasawa,A. (1981) *Tetrahedron Lett.,* **22**, 4755.
9. Gaffney,B.L. and Jones,R.A. (1982) *Tetrahedron Lett.,* **23**, 2257.
10. Trichtinger,T., Charubala,R. and Pfleiderer,W. (1983) *Tetrahedron Lett.,* **24**, 211.
11. Ti,G.S., Gaffney,B.L. and Jones,R.A. (1982) *J. Am. Chem. Soc.,* **104**, 1316.
12. Takaku,H., Morita,K. and Sumiuchi,T. (1983) *Chem. Lett.,* 1661.
13. Chaudhary,S.K. and Hernandez,O. (1979) *Tetrahedron Lett.,* **20**, 99.
14. Kuzmich,S., Marky,L.A. and Jones,R.A. (1983) *Nucleic Acids Res.,* **11**, 3393.
15. Gaffney,B.L., Marky,L.A. and Jones,R.A. (1984) *Tetrahedron,* **40**, 3.
16. Himmelsbach,F., Schulz,B.S., Trichtinger,T., Charubala,R. and Pfleiderer,W. (1981) *Tetrahedron,* **40**, 59.
17. Gao,X. and Jones,R.A. in preparation.

CHAPTER 3

Solid-phase Synthesis of Oligodeoxyribonucleotides by the Phosphite-triester Method

TOM ATKINSON and MICHAEL SMITH

1. INTRODUCTION

In recent years synthetic oligonucleotides have played a key role in answering some of the fundamental problems of molecular biology, for example in crystallographic and biophysical studies of RNA and DNA structure, as probes for isolation of cDNA and genomic DNA clones, as primers in RNA and DNA sequence determination and as site-specific mutagens (see Chapter 1).

Less than 5 years ago the synthesis of oligonucleotides was confined to a few specialist chemical laboratories. The preparation of protected nucleotide reagents and nucleoside-derivatised supports involved much time-consuming, skilful work. Today these reagents are now commercially available at reasonable cost. The introduction of solid-phase synthesis techniques and the possibility of automated synthesis has proved to be a powerful stimulus to streamline and simplify the chemical procedure. Today, the synthesis of oligonucleotides is no longer in the domain of specialist organic chemists. Molecular biologists can now rapidly assemble oligodeoxyribonucleotides by sequential addition of deoxyribonucleotide monomers using simple equipment, a set of the four deoxyribonucleoside-derivatised supports, the four deoxyribonucleotide monomers and simple effective procedures for the synthesis, deprotection and purification of oligonucleotides.

The first part of this chapter outlines the developments which led to stable monomeric deoxyribonucleotide phosphite reagents, and then describes in detail the preparation of deoxyribonucleoside-3′-phosphoramidites and of deoxyribonucleoside-derivatised supports. The second part gives detailed procedures for oligodeoxyribonucleotide synthesis, deprotection and purification, with notes on larger or smaller scale synthesis and multiple batch methods of synthesis. The last part lists commercial sources of reagents and apparatus.

2. THE CHEMISTRY OF SOLID-PHASE PHOSPHITE-TRIESTER SYNTHESIS

Oligodeoxyribonucleotides can be rapidly assembled by repetitive addition of deoxyribonucleotide monomers using solid-phase methods provided coupling efficiencies are consistently high. Synthesis on a solid support eliminates the need to

Phosphite-triester Method

isolate and purify the product after each chain extension step.

In solid-phase synthesis, the first deoxyribonucleoside is attached through its 3'-hydroxyl group to an inert insoluble support which also acts as a 3'-blocking device. The derivatised support is contained in a sintered glass funnel and reagents are allowed to react for specific times with the terminal deoxyribonucleoside and are then removed by filtration. The support is thoroughly washed with an appropriate solvent before the next reaction is carried out. In this way excess reagents are simply washed away and only the propagating oligodeoxyribonucleotide chains, anchored to the support, are retained within the reaction vessel.

Specifically, synthesis by the phosphite-triester method proceeds in the following steps. First, the 5'-hydroxyl groups of the deoxyribonucleoside attached to the support are deprotected. Next, they are condensed with excess activated 5'-O-dimethoxytrityl deoxyribonucleoside-3'-phosphoramidite solution. The new 3'-5' internucleotide phosphite-triester linkage is then oxidised to the more stable 3'-5' phosphotriester linkage. Any 5'-hydroxyl groups that failed to condense are capped as acetate esters. The dimethoxytrityl group, attached to the last nucleotide coupled, is removed to provide 5'-hydroxyl groups available for coupling with the next solution of activated deoxyribonucleotide monomer. The cycle is repeated until the oligodeoxyribonucleotide of desired sequence is constructed (*Figure 1*).

At the end of the synthesis the support carries a fully protected oligodeoxyribonucleotide. It is systematically deprotected and released from the support to yield the free oligodeoxyribonucleotide which can be purified by h.p.l.c. or by gel electrophoresis.

The development of this procedure was initiated in 1975 when Letsinger introduced a novel method of rapidly coupling two deoxyribonucleosides using reactive phosphite reagents (1). This new procedure had great potential for both oligoribo- and oligodeoxyribo-nucleotide synthesis. The method involved the reaction of a suitably protected nucleoside with a bi-functional phosphodichloridite to form a nucleoside-3'-phosphomonochloridite. This is reacted with a protected nucleoside. Mild oxidation generated the more stable 3'-5' internucleotide phosphotriester linkage. The procedure is summarised in *Figure 2*.

By 1981 Matteucci and Caruthers (2) had adapted this chemistry to solid-phase oligodeoxyribonucleotide synthesis using deoxyribonucloside-3'-phosphomonochloridites (X = Cl) or monotetrazolides (X = -N⟨N=N–N=N⟩) with deoxyribonucleoside-derivatised silica gels as the insoluble support (*Figure 3*).

The reactive deoxyribonucleoside-3'-O-phosphomonochloridites and phosphomonotetrazolides, despite their rapid and efficient coupling potentials, are not ideal reagents for oligodeoxyribonucleotide synthesis. They are exceptionally sensitive to hydrolysis and air oxidation, are difficult to prepare and isolate, require special handling techniques, have limited life times in solution and storage, and often contain appreciable amounts of an inert 3'-3' dinucleoside methox-

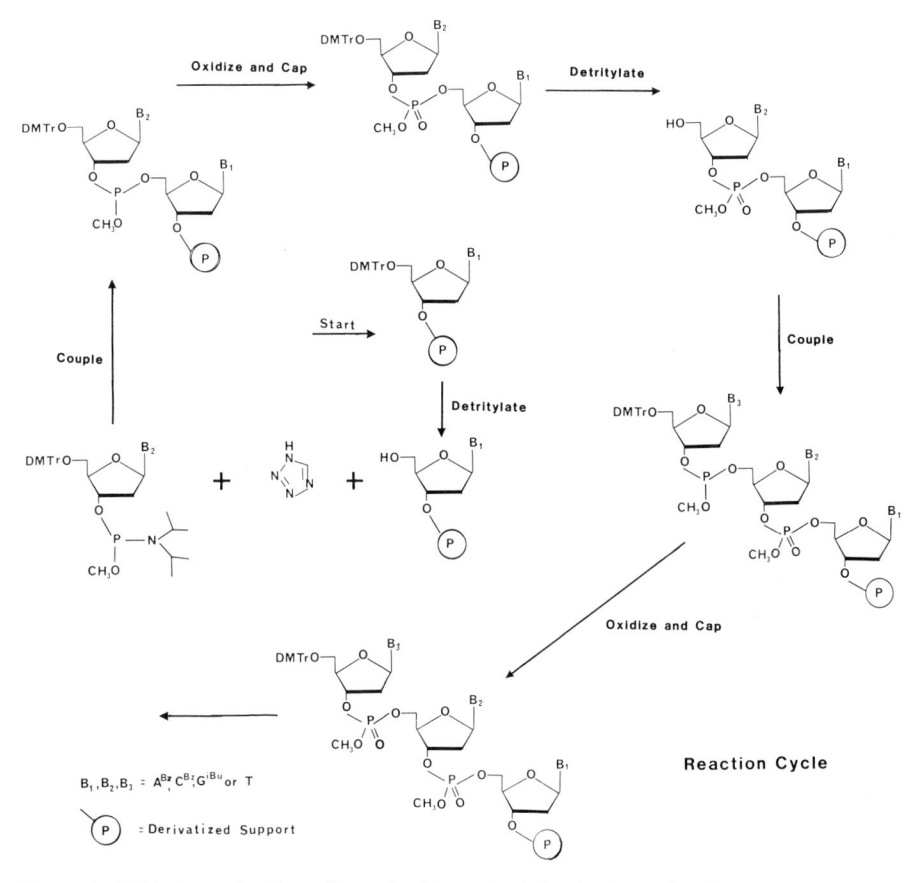

Figure 1. Solid-phase oligodeoxyribonucleotide synthesis by the phosphite-triester method.

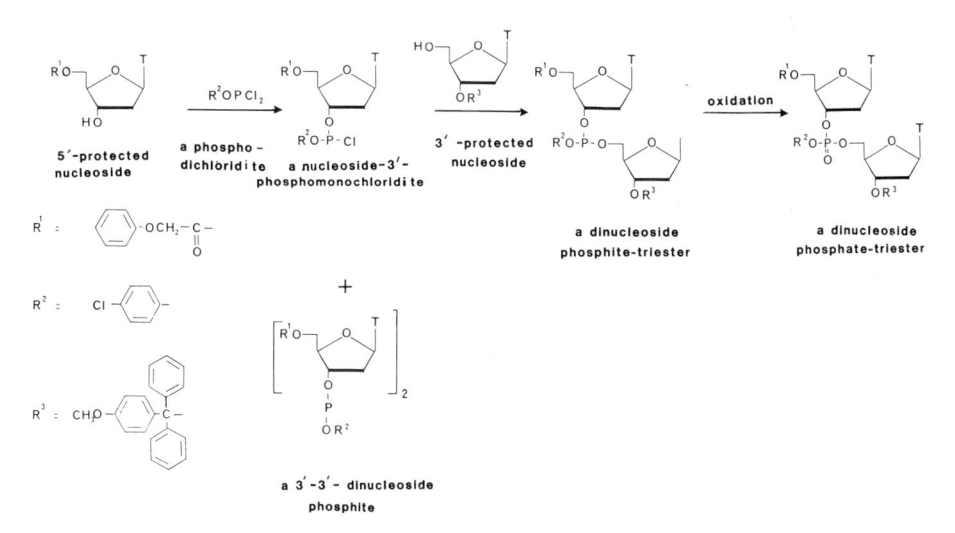

Figure 2. Letsinger's phosphite-triester method.

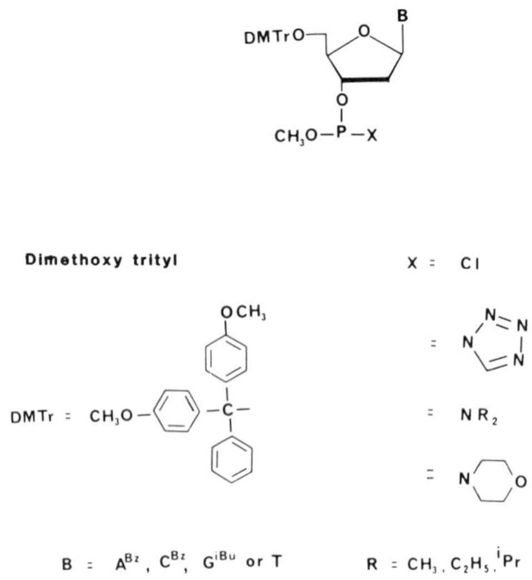

Phosphite-triester Method

Figure 3. Phosphite reagents introduced by Caruthers et al.

yphosphine. These problems were largely resolved when Beaucage and Caruthers (3) introduced a new class of deoxyribonucleoside phosphites: deoxyribonucleoside-3'-O(N,N-dimethylamino)phosphoramidites (*Figure 3*; X = NMe$_2$). These have improved stability and they are not hydrolysed by water or oxidised by air. Consequently they are much easier to prepare and use. Unlike the phosphomonochloridite and phosphomonotetrazolide derivatives, deoxyribonucleoside-3'-O-phosphoramidites cannot react directly with another 5'-hydroxyl-containing deoxyribonucleoside or deoxyribonucleotide. They must first be activated by treatment with a weak acid such as tetrazole.

In the synthesis of deoxyribonucleoside-3'-O(N,N-dimethylamino)phosphoramidites the use of a monofunctional phosphitylating agent, chloro-N,N-dimethylaminomethoxyphosphine (MeOPNMe$_2$), prevents the formation
$\quad\quad\quad\quad\quad\quad\quad\quad\quad\quad\quad\quad\quad\quad\quad\quad\quad\quad\quad\quad\quad\quad\quad\ \ \ |$
$\quad\quad\quad\quad\quad\quad\quad\quad\quad\quad\quad\quad\quad\quad\quad\quad\quad\quad\quad\quad\quad\quad\quad\ \ \ Cl$
of any 3'-3'-dinucleoside methoxyphosphine, but excess phosphitylating agent can cause variable amounts of deoxyribonucleoside-3'-O-phosphonate to be produced during work-up of the reaction. The phosphitylating reaction cannot be monitored, or the purity of isolated phosphoramidites estimated using t.l.c. because N,N-dimethylaminophosphoramidites decompose on the silica gel while the chromatogram is developing, even in solvent mixtures containing bases. Hence, the major contaminant in such preparations, deoxyribonucleoside-3'-O-phosphonate, cannot be removed by preparative silica gel chromatography, because of the lability of the phosphoramidites. The phosphonate is an acceptable contaminant because it is inert during oligonucleotide synthesis and its amount, in phosphoramidite preparations, can be estimated by ^{31}P n.m.r.

Solutions of deoxyribonucleoside-3'-O(N,N-dimethylamino)phosphoramidites in anhydrous acetonitrile display variable stability. In an attempt to discover

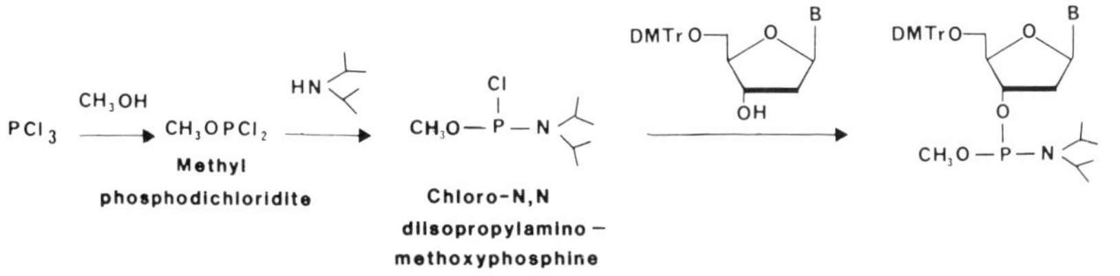

Figure 4. Monomer preparation.

more stable phosphoramidites, Adams *et al.* (4) prepared a series of N,N-dialkyl-aminophosphoramidites (*Figure 3*; X = NMe$_2$, NEt$_2$, N^{i}Pr$_2$) and demonstrated that increased steric hindrance about the nitrogen atom led to increased stability. McBride and Caruthers (5) confirmed that morpholino and diisopropyl phosphoramidites (*Figure 3*; X = N͡O, N^{i}Pr$_2$) are the reagents of choice for oligodeoxyribonucleotide synthesis. Both can be purified easily on a silica gel column, they are non-hygroscopic, and are stable indefinitely as dry powders at room temperature, and their solutions in anhydrous acetonitrile are stable at room temperature for at least a week. They have proved to be ideal reagents for oligodeoxyribonucleotide synthesis.

The commercial availability of pure, stable deoxyribonucleoside-3'-O-diisopropyl phosphoramidites and deoxyribonucleoside-derivatised solid supports has made oligodeoxyribonucleotide synthesis accessible to non-chemists. Molecular biologists can now rapidly synthesise an oligodeoxynucleotide of defined sequence by manual methods using simple equipment. A 20-mer can be assembled by sequential addition of deoxyribonucleotide monomers in less than a day and in the USA the cost per coupling can be less than $5 for all reagents.

Synthesis of deoxyribonucleoside-3'-O-phosphoramidites involves a two-stage preparation of phosphitylating reagent and then reaction with the 3'-hydroxyl group of a suitably protected deoxyribonucleoside (*Figure 4*). **Warning: the reactions used to prepare the phosphitylating reagent utilise hazardous chemisty and should only be performed by people competent in handling very reactive, moisture- and oxygen-sensitive, pyrophoric agents.**

Pure chloro-N,N-diisopropylaminomethoxyphosphine is commercially available and the preparation and purification of 5'-O-dimethoxytrityl deoxyribonucleoside-3'-O-diisopropyl phosphoramidites should present few problems to those familiar with organic synthesis. However, most people interested in using synthetic oligonucleotides will probably prefer to purchase pure deoxyribonucleoside-3'-O-phosphoramidites.

3. SYNTHESIS OF PROTECTED DEOXYRIBONUCLEOSIDE-3'-O-DIISOPROPYL PHOSPHORAMIDITES

3.1 Preparation of Methyl Phosphodichloridite

Methyl phosphodichloridite, MeOPCl$_2$, is prepared from phosphorus trichloride and methanol as described by Martin and Pizzolato (6).

Phosphite-triester Method

It is extremely important to use very dry reagents and apparatus. Glassware should be oven-baked at 120°C then cooled in a dessicator before assembly. Anhydrous methanol (BP 65.5°C) should be distilled from magnesium turnings and freshly distilled phosphorus trichloride (BP 76°C) should display only one ^{31}P n.m.r. signal. The reaction should be performed in an efficient fume hood, because copious quantities of hydrogen chloride gas are evolved.

(i) Add methanol (128 g, 157 ml, 4.0 mol) contained in a 250 ml pressure equalising addition funnel dropwise over 2–3 h to magnetically stirred phosphorus trichloride (550 g, 340 ml, 4.0 mol) at −20°C contained in a 1 litre three necked flask vented *via* an air condenser with a calcium chloride drying tube. During the addition maintain the temperature of the reaction between −20°C and −10°C with a dry ice/acetone bath.

(ii) Remove the bath when the addition is complete and allow the mixture to warm up to room temperature with stirring (~2 h).

(iii) Transfer the mixture to a dry 1 litre separating funnel closed with a calcium chloride drying tube and leave to stand overnight in the fume hood to de-gas.

(iv) Remove the lower layer, about 10–20 ml, and slowly add it to a large stirred volume of iced water.

(v) Fractionate the major phase by distillation at atmospheric pressure using a Vigreux column (300–400 mm long). The fraction boiling up to 85°C is mainly unreacted phosphorus trichloride. The fraction boiling between 85 and 93°C contains crude product and should be refractionated, collecting the fraction boiling between 92 and 93°C. If this fraction contains any phosphorus trichloride detected by ^{31}P n.m.r. it should be refractionated until it is pure by ^{31}P n.m.r. with a signal at −181 p.p.m. relative to 5% aqueous phosphoric acid v/v. Yields are variable, 20–50%.

Appreciable quantities of a red-orange polymeric residue containing elemental phosphorus are produced during the first fractionation. **Warning: it is important not to distil down to a small volume of residue, otherwise phosphorus may sublime or flash distil into the condenser and receiver flask.** The distillation apparatus should be left to cool down before dismantling it to reduce the risk of igniting the polymeric residue. The residue is destroyed by cooling the distillation flask in ice, adding solid carbon dioxide to provide an inert atmosphere and then, *very cautiously*, adding water.

The foreruns, containing largely unreacted phosphorus trichloride and some product, are destroyed by very slowly adding them to a large volume of stirred iced water. **Warning: too rapid an addition can lead to small explosions and fires. Methyl phosphodichloridite, crude or pure, reacts violently with water. Syringes or glassware coated with a film of it should be rinsed first with methanol before washing them with water.**

Methyl phosphodichloridite should be stored at −18°C.

3.2 Preparation of Chloro-N,N-diisopropylamino Methoxyphosphine

This reagent is prepared by the reaction of methyl phosphodichloridite with two equivalents of dry diisopropylamine.

(i) Dry dichloromethane (BP 40°C) by distillation from phosphorus pentoxide and diisopropylamine (BP 86°C) from calcium hydride. Dry all glassware in an oven as before.

(ii) Add diisopropylamine (101 g, 140 ml, 1.0 mol) in dry dichloromethane (80 ml), contained in a 250 ml addition funnel with a calcium chloride filled drying tube, dropwise over 1 h to a magnetically stirred solution of methyl phosphodichloridite (66.5 g, 47.1 ml, 0.5 mol) in dry dichloromethane (175 ml) in a 1 litre three necked flask at −10°C under an argon atmosphere.

(iii) Maintain the reaction temperature between −20° and −10°C during the addition using an acetone/dry ice bath.

(iv) Allow the stirred mixture to warm up to room temperature over 1−2 h.

(v) Remove diisopropylamine hydrochloride rapidly by vacuum filtration using dry apparatus and concentrate the filtrate. Dichloromethane is removed by distillation at atmospheric pressure or by evaporation under reduced pressure.

(vi) Fractionate the residue under reduced pressure. The fraction boiling between 83 and 86°C at 12 mm Hg should be pure product and display only one ^{31}P n.m.r. signal at −180.8 p.p.m. relative to a 5% aqueous phosphoric acid reference signal. If any methyl phosphodichloridite is detected, the sample should be refractionated until it is pure by ^{31}P n.m.r. Yield is 40−50%.

Warning: distillation foreruns should be cautiously disposed of by very slow addition to a large stirred volume of ice/water.

The pure phosphitylating reagent should be stored in small dry glass vials sealed with teflon-coated silicone-rubber serum caps inside a screw-capped glass jar containing self-indicating Drierite at −18°C in a freezer. The reagent must be allowed to warm up to room temperature before use and all transfers should be made *via* dry glass syringes under argon atmospheres.

3.3 Preparation of Deoxyribonucleoside-3'-O(N,N-diisopropylamino)phosphoramidites

Essentially this is the method described by McBride and Caruthers (5). The phosphitylating reaction is extremely moisture sensitive. Therefore all solvents, reagents and apparatus must be anhydrous.

(i) Dry diisopropylethylamine (BP 128°C) by distillation from calcium hydride.

(ii) Distil dichloromethane (BP 40°C) from phosphorus pentoxide and pass it down a short column of activated alumina just before use.

(iii) Dry the glass syringes used for injecting diisopropylethylamine, dichloromethane, phosphitylating reagent and methanol in an oven, then cool in a desiccator.

(iv) Allow the phosphitylating reagent (stored at −18°C) to warm to room temperature before use. Protect reagents from moisture and oxygen with argon-filled rubber balloons tied by elastic bands to 3 ml plastic syringe barrels with needles attached to capped serum bottles. This provides a convenient system for withdrawal or injection of liquids through the caps of the serum bottles (serum caps) under an inert atmosphere with pressure equalisation.

(v) Weigh sufficient of each 5'-O-dimethoxytrityldeoxyribonucleoside into a dry flask and co-evaporate twice with dry dichloromethane containing 10% (v/v) dry pyridine, and keep the residue under vacuum overnight. Even if the deoxyribonucleosides are known to be very dry, it is still recommended to keep them under vacuum overnight.

(vi) For a typical synthesis weigh deoxyribonucleoside (13.9 mmol) into a dry 100 ml round-bottomed flask.

(vii) Add a dry magnetic stirrer and cover the flask with a rubber serum cap (Aldrich Z10, 145-1 for size 24 flask necks).

(viii) Place an 18-gauge needle through the serum cap and the evacuate the flask overnight in a desiccator attached to a vacuum pump.

(ix) The next day slowly introduce argon into the desiccator, thus filling the flask.

(x) Place the flask on a magnetic stirrer, remove the venting needle and attach a needle to an argon-filled balloon placed through the serum cap. A similar needle, with argon-filled balloon, penetrates the serum cap of the vial containing the phosphitylating reagent.

(xi) Inject diisopropylethylamine (10.73 g, 12.1 ml, 4 x 13.9 mmol) onto the deoxyribonucleoside followed by dichloromethane (30 ml).

(xii) Stir the mixture. When all or most of the deoxyribonucleoside has dissolved inject the phosphitylating reagent (4.1 g, 4.0 ml, 1.5 x 13.9 mmol) over 20 sec.

(xiii) Stir the mixture for 15 min and monitor by t.l.c. This is done by withdrawing a small sample (0.1 ml) using a dry syringe and mixing it with water and ethyl acetate in a small tube. Apply a sample of the ethyl acetate (upper layer) to an aluminium-backed silica gel t.l.c. plate, develop in 45:45:10 v/v ethyl actate:dichloromethane:triethylamine and observe the plate under short-wave u.v. light. **Warning: wear appropriate eye protection**. Spray the plate with 20% sulphuric acid for the presence of dimethoxytrityl-bearing species (see Appendix I). *Figure 5* illustrates a typical t.l.c. plate and shows the four major components of the reaction mixture. All are u.v. absorbing and dimethoxytrityl bearing. The least polar (spot 1) is phosphoramidite and often appears to be dumb-bell shaped because the solvent system partially resolves the diastereoisomers due to the asymmetric phosphorus atom. Spot 2 is deoxyribonucleoside-3'-O-phosphonate and may result from hydrolysis of the phosphoramidite catalysed by hydrolysed excess phosphitylating reagent during the aqueous work-up of the reaction mixture. If this appears as the major component in the reaction mixture, the

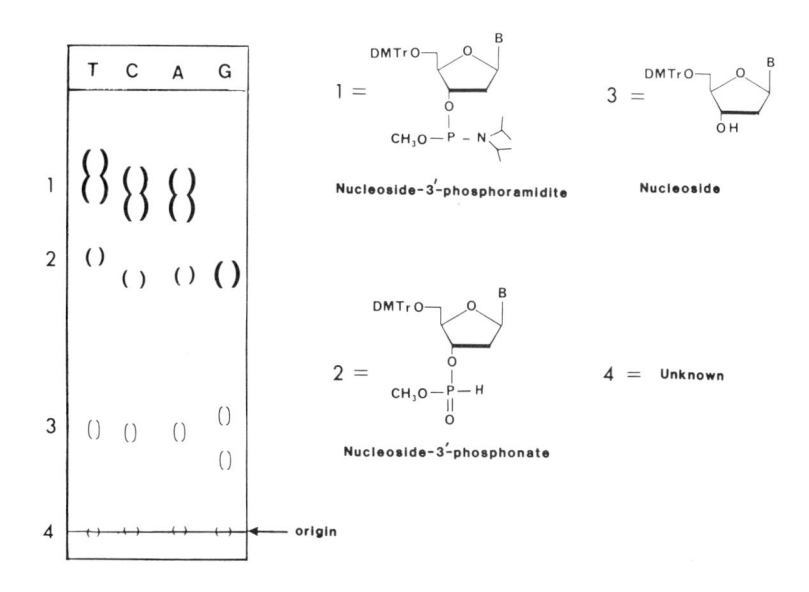

Figure 5. Monitoring the phosphitylating reaction by t.l.c.

reaction conditions were insufficiently anhydrous or the phosphitylating reagent has partially hydrolysed on storage. The last possibility can be checked by ^{31}P n.m.r. and the reagent redistilled if necessary. Spot 3 is starting deoxyribonucleoside and should not be observed in normal preparations. It is important that phosphoramidite preparations are devoid of deoxyribonucleoside. The deoxyribonucleoside will react with phosphoramidite and reduce the efficiency of oligonucleotide synthesis because phosphoramidite will be consumed by coupling with the contaminant deoxyribonucleoside as soon as activating tetrazole solution is added. If t.l.c. indicates that deoxyribonucleoside is present, more phosphitylating reagent should be added. Spot 4 is unknown polar material.

(xiv) When the reaction has gone to completion quench excess phosphitylating reagent by injecting anhydrous methanol (0.2 ml). This suppresses the formation of deoxyribonucleoside-3'-O-phosphonate during the aqueous washing steps to remove diisopropylethylamine hydrochloride.

(xv) Transfer the reaction mixture to a 1 litre separating funnel and dilute with ethyl acetate (300 ml) and triethylamine (15 ml), and then wash with 10% aqueous sodium carbonate (2 x 200 ml) and then with saturated aqueous sodium chloride (2 x 200 ml).

(xvi) Dry the organic phase over anhydrous sodium sulphate, filter, and then evaporate to a foam under reduced pressure.

(xvii) Deoxyribonucleoside-3'-O-phosphoramidites can be isolated by precipitating 35% solutions (the deoxyguanosine derivative in ethyl acetate and the other three derivatives in toluene) into rapidly stirred hexane (20 – 25 volumes). In the diisopropyl series the more soluble pyrimidine derivatives are precipitated at −20°C, whereas the purine phosphoramidites are effi-

Phosphite-triester Method

ciently precipitated at room temperature. In a typical preparation the crude reaction mixture will be more than 90% phosphoramidite. Since precipitation does not remove the more polar contaminants it is preferable to purify the phosphoramidites on a silica gel column using Dörper and Winnacker's method of column purification of morpholinophosphoramidites (7). Silica gel chromatography is fast and easy because the more polar by-products are retained on the column while the desired product elutes first.

(xviii) Purify 5 g portions of deoxyribonucleoside-3'-O-phosphoramidites on Kieselgel 60 (Merck Art 7734) contained in a glass column (20 x 5 cm). Chromatograph the more polar deoxyguanosine phosphoramidite on 80 g of silica gel and the other three phosphoramidites on 100 g of silica gel.

(xix) Use a freshly prepared solvent mixture consisting of 45:45:10 (by vol.) ethyl acetate:dichloromethane:triethylamine, colourless or freshly distilled, (BP 88.8°C) to pack the column, to dissolve the phosphoramidites, to elute the column and to develop the analytical t.l.c. plates.

(xx) Pack the column as a slurry, apply the sample (5 g in 10 ml solvent mixture) and elute the column at a flow-rate of ~ 8 ml per minute (see Appendix I). After applying the sample, elute about 140 ml before collecting 10 ml fractions.

(xxi) Collect about 30 fractions and analyse by t.l.c.

(xxii) Combine fractions containing pure product and then evaporate overnight under reduced pressure to remove traces of triethylamine. This results in a hard white foam or glass; store it in dry dark bottles in a desiccator at room temperature.

The yield of crude phosphoramidite usually exceeds 90% and recovery from the column is usually equally efficient.

Table 1. ^{31}P n.m.r. Chemical Shifts

PCl$_3$	−219
MeOPCl$_2$	−181
MeOP(Cl)N^{i}Pr$_2$	−180.8
DMTrdA^{Bz}p(OMe)N^{i}Pr$_2$	−148.67
	−146.47
DMTrdG^{iBu}p(OMe)N^{i}Pr$_2$	−148.47
	−148.33
DMTrdC^{Bz}p(OMe)N^{i}Pr$_2$	−148.74
	−148.33
DMTrdTp(OMe)N^{i}Pr$_2$	−148.94
	−148.54
DMTrdB	Not detected by ^{31}P n.m.r.

(B = A^{Bz}, G^{iBu}, C^{Bz} and T)

The ^{31}P n.m.r. chemical shifts relative to 5% aqueous phosphoric acid of species encountered in the preparation of deoxyribonucleoside phosphoramidites are given in *Table 1*.

4. PREPARATION OF SOLID SUPPORTS

The supports used in solid-phase synthesis should be inert towards, and insoluble in, all solvents and reagents used. They should be macroporous and have a large surface area to give rapid access to reagents and solvents. Supports should have a considerably larger particle size than the pores of the sintered glass frit used to retain the support and should possess good mechanical strength. They should also be easy to derivatise with sufficient and reproducible amounts of deoxyribonucleosides. Of the many types of solid support that have been investigated, silica or porous glass beads appear to best satisfy these criteria.

Silica gels of particle size less than 40 μm should not be used when sintered glass funnels of medium porosity (pore diameter 15 – 20 μm) are used as reaction vessels. 'Vydac' silica gel supports (particle size 20 μm) would not be effectively retained in such funnels, but derivatised 'Fractosil' (particle size 65 – 125 μm) and derivatised controlled pore glass (particle size 120 – 177 μm) are suitable supports for the filter funnel method of synthesis.

Fractosil consists of irregularly shaped silica particles which are brittle and occasionally fine fragments cause sinter blockage during a synthesis. This is a particular problem when small funnels (2 ml) are used.

Incomplete condensations using deoxyribonucleoside-derivatised aminopropyl silica gels are observed even when large excesses of activated phosphoramidite are used. Maximum coupling efficiencies are about 95 – 96% per condensation (8). This appears to be a steric effect because condensation using controlled pore glass supports with a longer linker attaching the deoxyribonucleoside to the support gives almost quantitative condensations (>98%) (4).

We prefer to use and recommend controlled pore glass supports because they do not cause sinter blockage problems even when using small funnels (2 ml capacity) and the more efficient couplings produce a greater yield of product and consequently a simpler purification using gel electrophoresis. The efficiency of assembly is estimated by comparing the dimethoxytrityl yield of the first deprotection with that of the last. If gel electrophoresis using the u.v. shadowing technique to detect oligonucleotides is used to isolate the desired oligonucleotide from truncated sequences an overall assembly efficiency of at least 10 – 20%, depending on oligonucleotide length, is required for a simple purification. Controlled pore glass supports are particularly recommended for synthesis of oligodeoxyribonucleotides containing more than 20 nucleotides.

Deoxyribonucleoside-derivatised solid supports of good quality are now commercially available. However, their preparation is straightforward. It involves the synthesis of the *p*-nitrophenyl esters of deoxyribonucleoside-3'-O-succinates and subsequent condensation with the primary amino groups of aminopropyl-derivatised silica gels, or long chain, alkylamino-derivatised, controlled pore glass (*Figures 6* and *7*).

Phosphite-triester Method

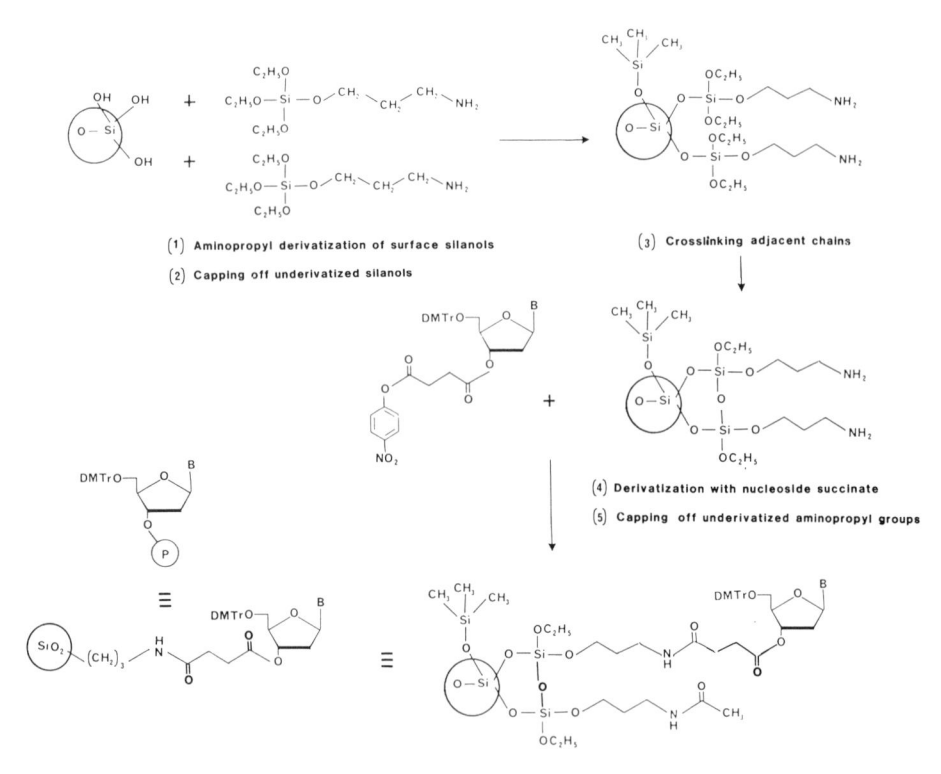

Figure 6. Preparation of deoxyribonucleoside-derivatised silica gels.

4.1 Preparation of Aminopropyl-derivatised Fractosil 500

(i) Suspend oven-dried Fractosil 500 (10 g particle size 63 – 125 μm; pore diameter 300 – 400 Å) **(Warning: it is essential that the oven is not being used for siliconising glassware while the silica gel is being dried otherwise surface silanol groups will be capped and the silica gel will be inert towards subsequent derivatisation)** in 95% ethanol (40 ml) in a conical flask.

(ii) To this add 3-aminopropyltriethoxysilane (10 ml) and swirl the mixture occasionally.

(iii) After 90 min at room temperature, filter the silica on a glass fritted funnel of medium porosity and wash with methanol until the effluent is no longer yellow, and then wash with ether and air dry.

(iv) Heat the silica for 1 h in an oven at 110°C and then cool in a desiccator. This causes cross-linking between adjacent 3-aminopropylsilylamino chains attached to the silica gel.

(v) Test a small portion for the presence of primary amino groups by the ninhydrin test (Appendix I). If the derivatisation is successful, the silica gel should be stained deep blue-purple, with very little colour in the supernatant. If the supernatant is appreciably coloured, excess 3-aminopropyltriethoxysilane is present due to inefficient washing of the silica gel, the methanol

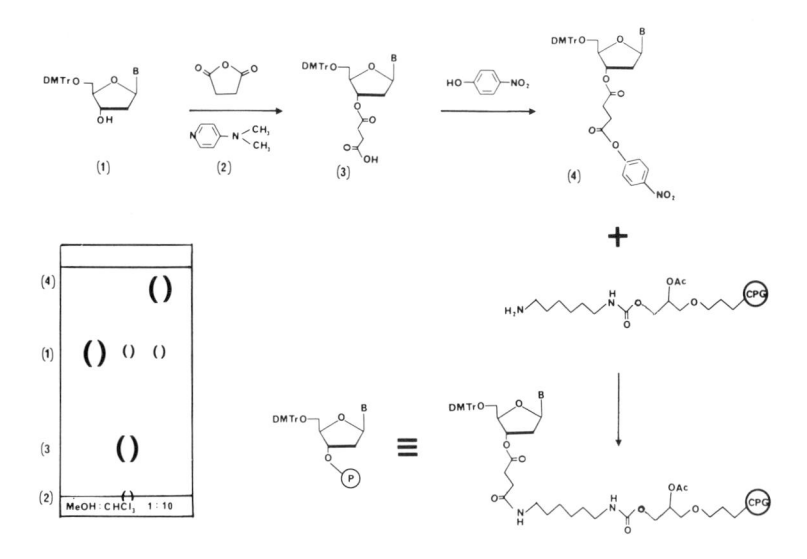

Figure 7. Preparation of deoxyribonucleoside-3′-O-succinates and deoxyribonucleoside-derivatised long chain alkylamine, controlled pore glass supports. For the sake of clarity, the capping of surface silanols and underivatised alkylamine chains is not illustrated.

and ether washes and the drying should be repeated. If the ninhydrin test gives a weak colour, the dry silica gel should be re-derivatised using a fresh bottle of 3-aminopropyltriethoxysilane. When the primary amino loading is satisfactory, underivatised hydroxyl groups on the silica gel are capped using trimethylchlorosilane.

(vi) Treat the support in a dry, stoppered flask, in a fume hood, with a clear solution of trimethylchlorosilane (10 ml) in dry pyridine (20 ml) for 2 h at room temperature. If an appreciable amount of precipitation occurs when the trimethylchlorosilane is added to the pyridine, the mixture should be discarded because a precipitate of pyridine hydrochloride indicates that the pyridine is not anhydrous. Another solution should be prepared using freshly distilled pyridine.

(vii) Filter the silica on a fritted glass funnel and wash extensively with methanol and ether and then dry in air.

(viii) Dry the support under reduced pressure. The aminopropyl-derivatised Fractosil is now ready for reaction with activated deoxyribonucleoside succinates.

4.2 Preparation of Deoxyribonucleoside-3′-O-succinates and the Deoxyribonucleoside-derivatised Silica Gels

(i) Add succinic anhydride (240 mg, 2.4 mmol) in portions over 30 min to a stirred solution of 5′-O-dimethoxytrityl N-acyl deoxyribonucleoside (3 mmol) in anhydrous pyridine (6 ml) containing 4-dimethylaminopyridine (180 mg, 1.5 mmol).

Phosphite-triester Method

(ii) Stir the reaction overnight and monitor by t.l.c. using chloroform:methanol 9:1 (v:v). The reaction is complete when very little deoxyribonucleoside is present (i.e., all the succinic anhydride has been consumed).
(iii) Evaporate the mixture under reduced pressure to a gum.
(iv) Remove residual pyridine by co-evaporation with dry toluene (3 x 20 ml).
(v) Dissolve the residue in dichloromethane (20 ml) and wash with ice-cold, 10% aqueous citric acid (2 x 15 ml) and then with water (2 x 15 ml).
(vi) Dry the organic phase over anhydrous sodium sulphate and evaporate under reduced pressure.
(vii) Dissolve the foam in dichloromethane (10 ml) and precipitate at room temperature into rapidly stirred hexane (250 ml).
(viii) Centrifuge the mixture, decant the supernatant carefully and dry the precipitate at the pump and then under vacuum. Yield is 75–85%.
(ix) Dissolve the succinylated deoxyribonucleoside (1 mmol) in dioxane (4 ml) containing dry pyridine (0.2 ml) and *p*-nitrophenol (140 mg, 1 mmol).
(x) Add dicyclohexylcarbodiimide (515 mg, 2.5 mmol). After a few minutes dicyclohexylurea begins to precipitate. Monitor the reaction by t.l.c., chloroform:methanol 9:1 (v/v) (*Figure 7*). It is usually complete within 2.5 h.
(xi) Remove dicyclohexylurea by filtation and add the supernatant to aminopropyl-derivatised Fractosil 500 (5 g) suspended in dimethylformamide (5 ml).
(xii) Add triethylamine (1 ml) and shake the mixture briefly by hand. A bright yellow colour, due to the release of *p*-nitrophenol, immediately develops. Swirl the reaction mixture occasionally and leave for 5 h or longer (overnight for convenience).
(xiii) Remove the supernatant and use it to derivatise another 2–3 g of aminopropyl Fractosil 500 because it still contains activated deoxyribonucleoside.
(xiv) Wash the deoxyribonucleoside-derivatised silica gel extensively with N,N-dimethylformamide, methanol and ether, dry in air and then *in vacuo*.
(xv) Assay the deoxyribonucleoside loading spectrophotometrically by determining the amount of dimethoxytrityl cation released by acidic treatment of a sample of the support. (a) Accurately weigh an aliquot of approximately 2 mg of 5′-O-dimethoxytrityl deoxyribonucleoside-derivatised support, then treat with perchloric acid solution (~4 ml) (70% $HClO_4$, 51.4 ml + MeOH 46 ml). (b) Measure the absorbance at 498 nm. If it is >1.8, dilute the sample and redetermine the absorbance. (c) The amount of bound deoxyribonucleoside in μmol/g of support, is given by:

$$\frac{\text{Absorbance at 498 nm x vol (ml) of } HClO_4 \text{ soln x 14.3}}{\text{wt of support (mg)}}$$

The support should have a loading of approximately 40–50 μmol of deoxyribonucleoside per gram. If the suport has a lower loading and free amino groups are still present (ninhydrin test) it can be re-derivatised using another sample of activated deoxyribonucleoside succinate. When the support has a satisfactory deoxyribonucleoside loading it is necessary to cap any underivatised amino groups with acetic anhydride to prevent side reactions which generate by-products during oligonucleotide synthesis.

(xvi) Treat the support (10 g) with acetic anhydride (2 ml) in dry pyridine (30 ml) containing 4-dimethylaminopyridine (100 mg) for 30 min.

(xvii) Filter the support, wash extensively with methanol and ether, air dry and then dry under vacuum in a desiccator. It should give a negative ninhydrin test and is ready for oligonucleotide synthesis.

4.3 Controlled Pore Glass Solid Support

Controlled pore glass consists of uniformly milled and screened particles of almost pure silica that are honeycombed with pores of a controlled size. It is manufactured from a borosilicate material that has been specially heat treated to separate the borates from the silicates. The pores are formed by removing the borates by an acidic etching process, their size is dependent on the nature of the heating procedure.

Beads of controlled pore glass (particle size 120–177 μm, pore diameter 500 Å) derivatised with a long chain alkylamine with a primary amino loading of about 100 μmol/g are available from Pierce Chemical Company. This can be reacted directly with 5'-O-dimethoxytrityl deoxyribonucleoside-3'-O-succinates using exactly the same method as that described for aminopropyl-derivatised silica gel. Thus, a solution containing 1 mmol of the p-nitrophenyl ester of deoxyribonucleoside-3'-O-succinate will derivatise 5 g of long chain alkylamine, controlled pore glass to give a support with a loading of about 25–30 μmol of deoxyribonucleoside per gram (4) (*Figure 7*). This support has a much greater mechanical strength compared with the Fractosil based supports and it does not block the sintered glass filter because of fragmentation. In addition, it gives faster and more efficient couplings than do derivatised aminopropyl silica gels.

5. ASSEMBLY OF PROTECTED OLIGODEOXYRIBONUCLEOTIDES
5.1 Reactions Involved in the Assembly Cycle
5.1.1 *Removal of 5'-O-Dimethoxytrityl Group*

The dimethoxytrityl group, being acid-labile, is conveniently removed using 3% dichloroacetic acid (pK_a 1.5) in dichloromethane (4). N-benzoyldeoxyadenosine is also acid-labile and some depurination may occur on exposure to acidic solutions. The risk of depurination is greatest when the N-benzoyldeoxyadenosine residue is at the 5' end of the propagating chain (8). Internal N-benzoyldeoxyadenosine residues that are flanked by 5'- and 3'-phosphates are stable under acidic detritylation conditions. Of the large number of reagents that have been investigated, dichloroacetic acid in dichloromethane is the most effective in achieving rapid detritylation with minimum loss of N-benzoyladenine.

The purine deoxyribonucleosides detritylate faster than the pyrimidines, thus 5'-deoxyadenosine residues require the shortest exposure to acid and deoxycytidine residues the longest. It is extremely important not to leave the support in contact with the acid for longer than the times specified.

Phosphite-triester Method

Acidic solutions containing the dimethoxytrityl cation are bright orange. The effluent solution after each detritylation step can be collected and assayed spectrophotometrically to give the coupling efficiency of each condensation. It is worthwhile doing this for the first few syntheses to ensure the method is working adequately and that each of the deoxyribonucleotide monomers is coupling efficiently. The first and last detritylation solutions of all syntheses should be collected to determine the assembly efficiency of the oligonucleotide in order to estimate how much of the crude mixture to load on a gel to give about $0.5-1.0\ A_{260}$ units of pure oligonucleotide. The support is washed with dichloromethane before detritylation to remove any chemical species that may quench the trityl cation, and after the detritylation reaction with dichloromethane to ensure complete removal of dichloroacetic acid from the support.

Warning: dichloroacetic acid can cause burns to skin or clothing. If contaminated, wash with copious amounts of water. Dichloromethane should always be used in a hood.

5.1.2 *Coupling of Deoxyribonucleoside Phosphoramidite to the 5' End of Oligonucleotide Attached to Solid Support*

The coupling reaction is sensitive to air and moisture. The support is dried under vacuum and an argon atmosphere introduced before the activated deoxyribonucleotide monomer is added. Deoxyribonucleoside-3'-O-phosphoramidites require activation to react with the 5'-hydroxyl groups of the support-bound oligonucleotide. The phosphoramidites are activated by tetrazole, which acts by protonating the nitrogen atom of the phosphoramidite (3) and also by forming a deoxyribonucleoside-3'-O-phosphomonotetrazolide (5) *via* a substitution reaction. Both of these species undergo rapid nucleophilic substitution reactions with the hydroxyl groups of support-bound oligonucleotides to form an internucleotide phosphite-triester linkage. In the procedure used here, an aliquot of the appropriate phosphoramidite solution is mixed with an aliquot of tetrazole solution in a dry, argon-filled Eppendorf tube and then injected immediately onto the support, shaken gently and then allowed to react for 3 min. The coupling mixture is removed by filtration and the support washed with acetonitrile before the new internucleotide phosphite-triester linkage is oxidised to the more stable phosphotriester linkage.

5.1.3 *Oxidation of the Phosphite-triester to Phosphotriester*

Oxidation of phosphite to phosphate is complete in less than 1 min using a mixture containing iodine in tetrahydrofuran, 2,6-lutidine and water. The support is washed well with acetonitrile until it and the effluent are colourless before the capping reagent is added.

5.1.4 *Capping to Block Unreacted 5'-Hydroxyl Groups*

A capping step is introduced into the reaction cycle after the coupling and oxidation steps to block any 5'-hydroxyl groups that failed to condense with activated phosphoramidite solution. This ensures that the subsequent reactions proceed only by propagating chains of the desired sequence. It gives a much higher yield of product and a much easier purification at the end of the synthesis. The 5′-hydroxyl functions are acetylated by acetic anhydride rendering them inert towards further chain extension. 4-Dimethylaminopyridine (DMAP) catalyses the acetylation reaction (9). Aliquots of acetic anhydride and DMAP solutions are mixed just before use in every cycle because the reaction mixture darkens and deteriorates rapidly. Capping is complete within 2 min. Excess reagent is removed by filtration and the support washed with acetonitrile and then with dichloromethane before the start of the next cycle of synthesis.

The inclusion of a capping step gives two further advantages. Since only one species is being chain-extended the quantity of dimethoxytrityl cation released (by the detritylation reaction) is directly proportional to the efficiency of the last coupling reaction. A collection of solutions of dimethoxytrityl cation can be assayed to give the individual coupling efficiencies and also the overall assembly efficiency of support-bound oligonucleotides. The excess capping mixture also scavenges any residual water left on the support by the aqueous oxidising mixture.

5.2 Reagents and Solvents

The reagents and solvents used in solid-phase oligodeoxyribonucleotide synthesis should be of the highest purity in order to avoid contaminants which cause undesired by-products and inefficient reactions. H.p.l.c. grade solvents are recommended. This is not extravagant because reagents and solvents are used in small amounts and account for less than 10% of the cost of an oligonucleotide; the bulk of the cost is expended on the phosphoramidites.

The reagent solutions for the coupling reaction, the tetrazole and phosphoramidite solutions, are prepared on the day of synthesis in small dry serum bottles blanketed by argon atmospheres. These solutions are stable for at least 2−3 days and should give good couplings as long as they remain anhydrous.

The tetrazole-activated phosphoramidite solution and the capping mixture are prepared just before use in each cycle by mixing the correct amounts of the two components of the reagent mixture.

The detritylation, oxidation, deprotection solutions and both components of the capping mixture are stable for extended periods (several months) when stored at room temperature in dark glass, screw-capped bottles, but it is good practice to make up sufficient stock solution for about 100 couplings so that these solutions are prepared fresh every 2−3 weeks.

The composition of these stock solutions, sufficient for 100 couplings when using 1.5 μmol of deoxyribonucleoside-derivatised support, is given below. Sources of reagents and solvents are given in Section 9.

Detritylation: 3% (v/v) DCA: 9 ml dichloroacetic acid + 291 ml dichloromethane. Use about 4 ml per cycle.

Coupling: (i) 0.5 M tetrazole solution: 175 mg sublimed tetrazole in 5 ml anhydrous acetonitrile. Use 0.18 ml per coupling (90 μmol).

	(ii) 0.2 M 5′-O-dimethoxytrityl deoxyribonucleoside-3′-O-phosphoramidites in anydrous acetonitrile. Use 0.15 ml per coupling (30 μmol).
Oxidation:	0.1 M iodine solution: 2.6 g iodine + 80 ml tetrahydrofuran (THF) + 20 ml 2,6-lutidine + 2.0 ml water. Use about 1 ml per cycle.
Capping:	(i) 6.5% (w/v) DMAP solution: 6.5 g 4-dimethylaminopyridine + 100 ml anhydrous THF. Use 0.75 ml per cycle. (ii) acetic anhydride/2,6-lutidine: 10 ml acetic anhydride + 15 ml 2,6-lutidine. Use 0.25 ml per cycle.

Deprotection of methyl ester with triethylammonium thiophenate:

NEt_3H^+ PhS^- solution: 4 ml thiophenol + 8 ml triethylamine + 8 ml dioxane. **Warning: toxic stench — store and use in fume hood.** Use 1 ml per synthesis. **Wear gloves.**

Concentrated ammonia solution **Stored at 0°C tightly capped**

Methanol h.p.l.c. grade

Solvent for washes:

Acetonitrile	h.p.l.c. grade
Dichloromethane	h.p.l.c. grade

Anhydrous solvents and reagents that require distillation from calcium hydride:

Acetonitrile	h.p.l.c. grade	for tetrazole and phosphoramidite solutions.
THF	h.p.l.c. grade	for DMAP solution
2,6-lutidine		for acetic ahydride/2,6-lutidine mixture

The deoxyribonucleoside-derivatised supports should be stored in tightly capped, dark glass bottles sealed with Parafilm in a desiccator over indicating Drierite at room temperature.

The deoxyribonucleoside diisopropyl phosphoramidites should be stored as dry powders in a similar manner. They can be handled like any other stable organic solid, but be sure that there are no acidic vapours present when the bottles are opened.

Solid tetrazole can be stored in tightly closed, dark glass bottles in a cupboard at room temperature. Do not use the tetrazole if it smells of acetic acid; use another bottle. Use only sublimed tetrazole. This should be microcrystalline and totally dissolve to give 0.5 M solutions. Once prepared, the DMAP and iodine stock solutions should be filtered through paper to remove small amounts of insoluble material.

Concentrated ammonia solution contains about 30% NH_4OH. Solutions stored at room temperature become progressively less concentrated because of loss of ammonia gas. Consequently the ammonia solution should be stored in a tightly capped bottle in a cold room otherwise loss of ammonia will decrease the effectiveness of the solution as a deprotecting reagent.

5.2.1 Distillation of Solvents Using a Continuous Reflux Apparatus

Anhydrous acetonitrile is required for the coupling reaction and anhydrous tetrahydrofuran is required for the capping reagent. These solvents can be dried by distillation from a desiccant (calcium hydride). Pyridine, required for the phosphotriester method (Chapter 4) can be distilled from potassium hydroxide by this method or that given in Appendix I.

Distillation using a continuous reflux apparatus has several advantages compared with a conventional distillation set up:

(i) the apparatus is simpler and easier to assemble;
(ii) its compact vertical design requires very little bench space;
(iii) the unit can be left unattended with no risk of the distillation flask boiling dry;
(iv) the unit need not be disassembled when not in use.

The apparatus (*Figure 8*) consists of a head storage unit which is placed between the distillation flask and a 250 mm reflux condenser, which is capped by a drying tube containing indicating Drierite. The head storage unit can either collect distillate (excess of 250 ml is returned to the distillation flask *via* the overflow tube) or return distillate to the distillation flask, *via* the double oblique stopcock, or to a dry collection flask.

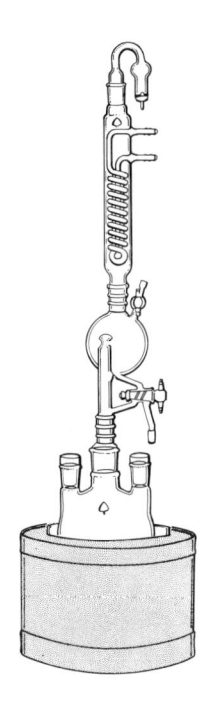

Figure 8. A continuous reflux apparatus for distillation of solvents.

Phosphite-triester Method

Procedure

Acetonitrile boils at 81 – 82°C.

Tetrahydrofuran boils at 65.5 – 66.6°C.

Pyridine boils at 115°C.

(i) All glassware should be oven-baked at >110°C for at least 2 h to dry before use. (The teflon stopcocks should be removed from the head storage unit.)
(ii) Assemble the apparatus as in the diagram, in a fume hood, remote from open flames.
(iii) Add 3 or 4 anti-bumping granules (Boileezers – Fisher).
(iv) Add the appropriate drying agent ~15 – 20 g.
(v) Add no more than 500 – 600 ml of solvent (preferably the most anhydrous h.p.l.c. grade available).
(vi) Make sure water is flowing through the condenser.
(vii) Set the oblique double stopcock to drain distilled solvent to the distillation flask.
(viii) Bring the solvent to gentle reflux.
(ix) Reflux for about 60 min. (This further dries solvent and glassware.)
(x) Set the oblique stopcock to collect solvent in the storage unit.
(xi) Collect about 80 ml as forerun, discard this.
(xii) Collect 250 ml in a dry flask, and the next 150 ml. (*Never* distil to dryness.)
(xiii) Allow apparatus to cool.
(xiv) Add more solvent if required and repeat steps (vi – xiv). The desiccant should be removed after every two or three still refills.
(xv) Discard spent desiccant carefully. Decant residual solvent from the desiccant and discard. Potassium hydroxide may be disposed of by dissolving in about 700 ml water. This solution should be neutralised with glacial acetic acid or hydrochloridic acid and then poured down the drain. Residual calcium hydride should be placed in a dry beaker and exposed to atmospheric humidity in a fume hood for 3 – 4 days. During this time most of the grey granules should have broken down to an amorphous grey-white powder. This can be destroyed by carefully spooning out small portions and adding them to a half-filled 2 litre beaker of water. Effervescence indicates some calcium hydride is still present. Add 30 – 40 ml of glacial acetic acid or hydrochloric acid to dissolve the grey sludge of calcium hydroxide and then pour the solution down the drain, diluted with plenty of water.

5.3 Apparatus

The apparatus is shown in *Figure 9*. Liquids are retained by surface tension within medium porosity (pore size 10 – 15 μm) glass filter funnels and can be removed by the application of gentle vacuum. Such funnels may be used as simple inexpensive solid-phase reaction vessels. Reagents are allowed to sit in contact with the immobilised oligonucleotide chains for specific times and are then removed by filtration.

Glass funnels of 2 ml capacity are suitable for small scale syntheses using 40 – 60 mg of deoxyribonucleoside-derivatised controlled pore glass supports

T. Atkinson and M. Smith

(1) Washing support

Trityl collection flask

Water pump

Organic waste flask

(2) Introducing argon atmosphere

Argon

Water pump

Coupling flask

(3) Coupling

Coupling flask

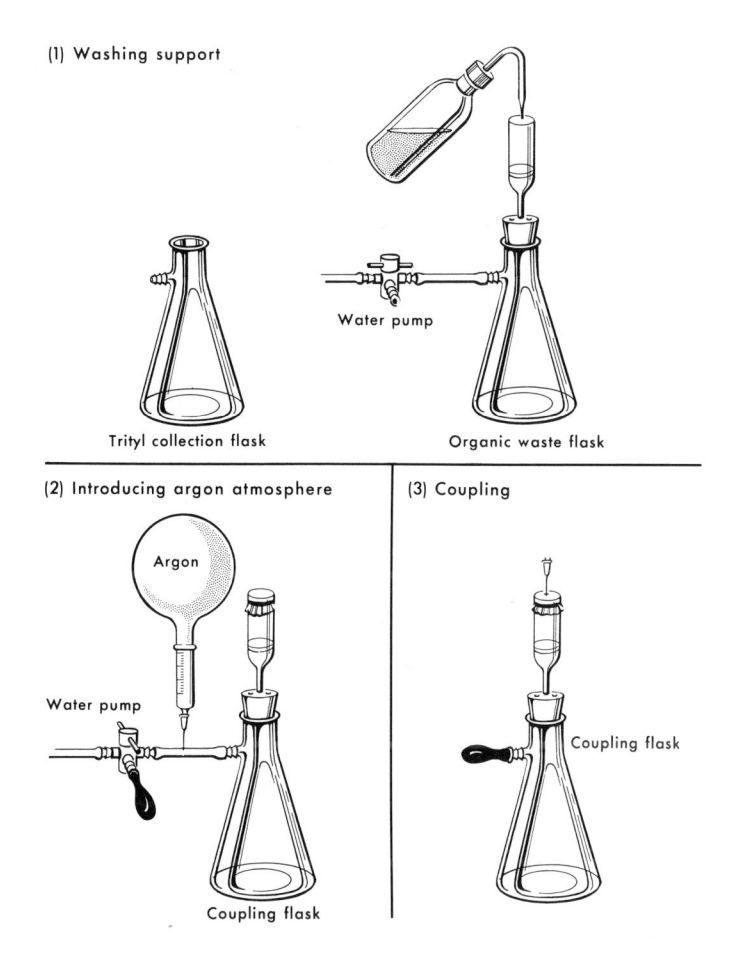

Figure 9. Oligodeoxyribonucleotide synthesis apparatus.

(i.e., 1.0 – 1.5 μmol of support-bound deoxyribonucleoside), whereas funnels of 15 ml capacity are suitable for syntheses using up to 5 μmol of bound deoxyribonucleoside and for all syntheses involving deoxyribonucleoside-derivatised Fractosil supports.

Dry the funnels in an oven and then siliconise using 1% trimethyl chlorosilane in toluene or chloroform before use. Seat the funnel in a rubber stopper (size 5) attached to a filter flask (125 ml) secured by a clamp and connected to a vacuum line. Collect all organic waste solutions except the detritylation solutions in this flask, which should be periodically emptied into a bottle for organic waste solvents. Collect the orange dimethoxytrityl solutions together with the dichloroacetic acid washes in a similarly clamped flask. Move the funnel and vacuum line from the detritylation flask to the organic waste flask and then back when appropriate. Transfer the contents of the dimethoxytrityl flask to a test tube and assay the series of dimethoxytrityl solutions spectrophotometrically for individual coupling efficiencies.

Phosphite-triester Method

Use a third flask only during the support drying step and during the coupling reaction. Nothing is filtered into the coupling flask, which is stored in a desiccator during the rest of the cycle.

'Twistfit' brand stoppers have twist off seals and can be used as zero, one, two or three holed stoppers, this avoids the inconvenience of having to bore holes in stoppers. The stem of the funnel will slide more easily into the hole in the stopper if it is lubricated with a little ethanol.

A 'Nalgene' three-way stopcock (4 mm bore) placed in the vacuum line to a water aspirator gives a simple method of controlling the vacuum. Covering the open limb with a finger will complete the vacuum to the filter flask and drain the funnel of liquid. Lifting the finger from the port will release the vacuum.

Rubber sleeve-type serum caps of three sizes are used to cover 1.5 ml Eppendorf tubes, serum bottles and filter funnels. In order to avoid unnecessary repetition of descriptions and catalogue numbers they will be referred to as small, medium or large. Small serum caps (stopper size: 7 mm) (Aldrich Z10, 073-0 or Sigma S5634) are used to cover lidless Eppendorf tubes and 2 ml capacity filter funnels. Medium sized serum caps (stopper size: 13 mm) (Aldrich Z10, 074-9 or Sigma S0260) are used to cover serum bottles and large serum caps (stopper size: 24 mm) (Aldrich Z10, 145-1 or Sigma S501) are used to cover 15 ml capacity filter funnels.

The tetrazole and phosphoramidite solutions in anhydrous acetonitrile are contained in 5 ml dark glass serum bottles (Aldrich Z11, 402-2) sealed by medium sized serum caps. Inert, anhydrous atmospheres over these solutions are provided by small argon-filled balloons tied by elastic bands to 3 ml plastic syringe barrels with 1 inch long 23-gauge needles (Becton Dickinson B-D 5571) which penetrate the serum caps. The balloons provide a simple method of pressure equalisation while withdrawing and delivering anhydrous solutions through serum caps by syringe. When a serum cap is over-punctured, by repeated use, the balloon will deflate and the serum cap should be replaced.

Attach a large argon-filled balloon to a 10 ml plastic syringe barrel with an 18-gauge needle and use it to fill the coupling flask and serum capped funnel with argon just before the activated phosphoramidite solution is injected onto the support.

Inflate the balloons by placing the needles through a clamped off length of rubber tubing attached to a cylinder of anhydrous oxygen-free argon. Once a balloon is inflated, insert the needle into a rubber stopper to prevent loss of gas, and store in this manner until required.

The serum bottles containing the anhydrous tetrazole and phosphoramidite solutions are top heavy when balloons are attached. They can be supported and organised in a rack made from a styrofoam block by cutting out suitably sized holes with cork borers (see *Figure 10*).

Mix aliquots of tetrazole and phosphoramidite solutions in dry tubes, prepared by covering lidless Eppendorf tubes (1.5 ml) with small serum caps, inserting 1 inch long disposable 23-gauge needles through each serum cap and keep the tubes overnight in a desiccator under vacuum. Bleed argon into the desiccator on the day of synthesis, filling the tubes *via* the venting needles.

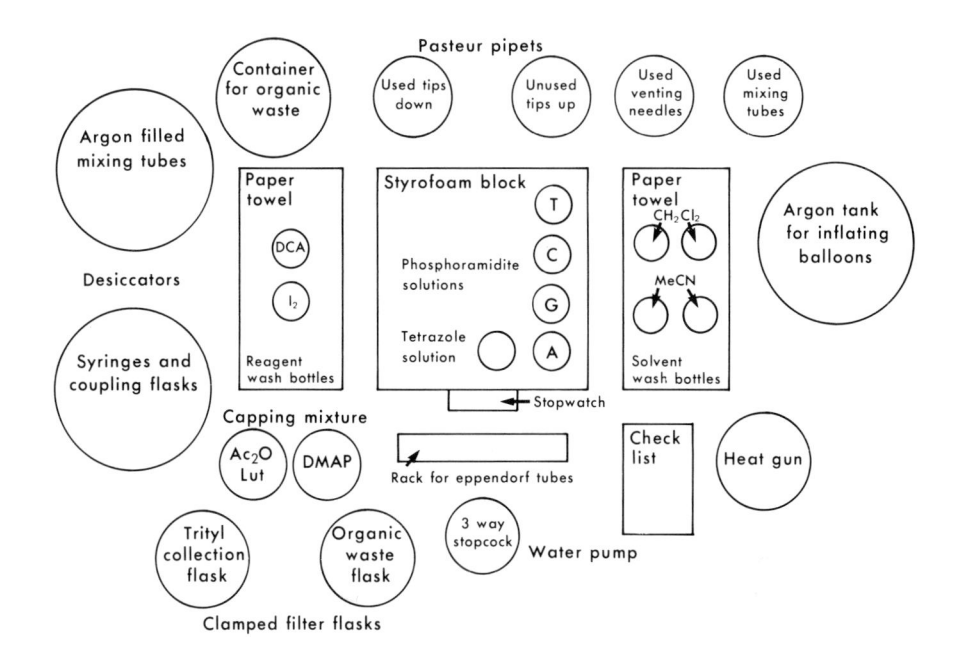

Figure 10. Layout of synthesis apparatus.

Freshly distilled anhydrous acetonitrile, used for dissolving tetrazole and the phosphoramidites, is contained in a 30 ml serum bottle (Aldrich Z11, 398-0) sealed by a medium sized serum cap and is under an argon atmosphere.

Use glass Tuberculin syringes (1/4 ml or 1 ml; B-D 2001 and B-D 2004, respectively) to transfer tetrazole and phosphoramidite solutions to the mixing tube and then to transfer the activated phosphoramidite solution to the support.

Assign an oven-dried, desiccator-cooled, syringe to each serum bottle: one for tetrazole, one for anhydrous acetonitrile and one for each of the phosphoramidites. Label each syringe and colour code to prevent cross-contamination of syringes and solutions, and fit with a 2 inch long stainless steel, 20-gauge Luer Lok hub needle (B-D 1078). Needles of this length are required to retrieve all the solution from the serum bottles. Needles of finer bore such as 22-gauge are too pliable for convenience, while needles of large bore (18-gauge) tend to remove part of the serum caps.

During the synthesis store the syringes in a desiccator containing indicating Drierite, remove as required, and then return to the desiccator immediately after use. It is unnecessary to wash out and dry the syringes after each operation provided the plunger moves freely within the syringe barrel. If washing is necessary, wash well with acetone and then oven dry for at least 5 min, before returning it to the desiccator. When the synthesis is complete, or at the end of a day of synthesis, the syringes should be washed well with acetone and then baked in an oven. Care must be taken not to interchange syringe barrels and plungers. They are usually numbered and matched for an airtight fit.

Phosphite-triester Method

Detritylating reagent, oxidising reagent and undistilled acetonitrile and dichloromethane are contained in Nalgene wash bottles (60 ml). The wash bottles are used to squirt reagents and solvent directly onto the support. It is a good idea to segregate the two reagent wash bottles from the solvent wash bottles in order to minimise the risk of accidentally adding dichloroacetic acid solution or iodine solution instead of a solvent wash or *vice versa*. The wash bottles should be stood on paper towels to absorb any drips or spillage.

The two component solutions of the capping mixture are stored in small dark screw-capped bottles (25 ml capacity). Mix aliquots of the solutions in an Eppendorf tube or a small glass tube and then transfer to the support by Pasteur pipette.

A Savant 'Speed Vac Concentrator' is a small-scale, rotary evaporator capable of evaporating the contents of several Eppendorf tubes (up to 40) simultaneously. Such a device is required to remove ammonia solution when the deprotection of the oligonucleotide is complete and also to evaporate aqueous solutions of oligonucleotide before storage.

5.4 Procedure for Oligodeoxyribonucleotide Synthesis

It is convenient to carry out synthesis using 1.5 μmol of support- bound deoxyribonucleoside (i.e., ~60 mg of a controlled pore glass support having a loading of 25 μmol/g) contained in a 2 ml siliconised glass filter funnel of medium porosity.

A typical synthesis of 15 – 20 nucleotides in length should yield between 60 and 120 A_{260} units of crude oligomer. Purification of 10 – 15% (i.e., ~10 A_{260} units) of crude mixture by gel electrophoresis should produce about 0.5 – 2.0 A_{260} units of pure oligomer. This is more than adequate for most purposes.

The synthetic cycle consists of four fast reactions: capping, detritylation, coupling and oxidation, with the appropriate solvent washes between reactions to remove excess reagents from the support and from the funnel walls. During reactions a reagent solution is added, the support is agitated gently and then allowed to react for a specific time.

During the solvent washes agitate the support by squirting the solvent directly onto it. Remove the solvent by gentle filtration (partial vacuum) and repeat the wash if necessary. **When reagents and solvents are removed by filtration, it is absolutely essential to avoid pulling air through the support — this is particularly important on humid days.** The liquid level should be brought down to the surface of the support, the vacuum released and the next wash or reagent added. It is important not to leave the reagents in contact with the support for longer than the times specified, particularly during the acid-catalysed detritylation reaction. It is safe to interrupt the synthesis for a short time (a few minutes) after a washing step has been completed. The safest place to interrupt a synthesis during the day or at the end of a day is just before the coupling reaction (step 9) when the support is dry and covered with argon. Cover the funnel with a serum cap and attach it to the argon-filled coupling flask, the side arm of which should be sealed with a rubber pipette tip. Leave the funnel and flask in a desiccator at room temperature.

To resume synthesis repeat the drying steps and then continue with the next coupling.

When the last phosphoramidite of the synthesis has been coupled and oxidised, omit the capping step. Wash the support with acetonitrile and dichloromethane, detritylate, wash again with dichloromethane and air dry. The synthesis is now complete and the oligonucleotide is ready for deprotection.

The support should remain in the funnel if the thiophenol treatment is to be used but transferred to an Eppendorf tube (1.5 ml) if the cold and hot ammonia deprotection scheme is to be used.

Synthesis should be performed in an area of the laboratory in a hood remote from additional sources of humidity such as water baths, autoclaves and running hot water taps. Synthesis commences by treating the support with capping mixture to ensure there are no free primary amino groups on the support which could interfere with the coupling reaction. Transfer an aliquot of acetic anhydride in lutidine (0.25 ml) to an Eppendorf tube, which has been previously calibrated to 1.0 ml by a P1000 Pipetman, and transfer sufficient DMAP solution by Pasteur pipette to make the volume of the mixture up to the 1 ml mark on the tube. Use the Pasteur pipette to transfer the mixture to the support and then discard the pipette. Remove the capping mixture 2 min later by filtration and wash the support with acetonitrile (2 x 2 ml) and then with dichloromethane (2 x 1.5 ml).

Transfer the funnel and vacuum line to the dimethoxytrityl collection flask. Squirt about 1 ml of DCA solution onto the support with gentle agitation. The characteristic orange colour, due to the released dimethoxytrityl cation, immediately develops. Expose $5'$-deoxyadenosine residues to two 20 sec treatments of DCA solution (1 ml each) followed by two 0.5 ml washes of DCA solution, while the other residues should have two 30 sec treatments of 1 ml followed by 3 or 4 x 0.5 ml washes until the support and effluent are colourless. Transfer the contents of the coupling flask later in the cycle to a serum-capped test tube marked with the number of the coupling and 'deoxyribonucleotide just detritylated'. Observation of a collection of such tubes gives a rapid qualitative check on coupling efficiencies. These solutions can be assayed later to calculate individual coupling efficiencies.

Return the funnel and vacuum line to the organic waste filter flask. Wash the support and funnel walls with dichloromethane (5 x 2 ml). The nozzle of the dichloromethane wash bottle should be wiped with a paper tissue before use to remove any water or ice that may have condensed from the air, owing to the evaporation of droplets of dichloromethane on the end of the nozzle. If this proves to be a serious problem, this wash should be replaced by an anhydrous acetonitrile wash (5 x 2 ml). The anhydrous acetonitrile should be stored in 30 ml serum bottles (sufficient for three cycles) and closed by a serum cap. The dry acetonitrile should be poured directly into the funnel from the open serum bottle. The serum cap should be promptly replaced on the serum bottle after the wash.

When the dichloromethane (or anhydrous acetonitrile) wash is complete transfer the funnel to the coupling flask and cover with a small serum cap. Cover the open vent of the three-way stopcock by a rubber pipette bulb and expose the system to full vacuum. Heat the funnel gently with a hair dryer or heat gun on a

Phosphite-triester Method

low setting. The funnel should be warm to the touch, but not hot. After about 1 – 2 min when the support is dry and mobile and all the solvent has evaporated, turn the stopcock to isolate the funnel and flask from the vacuum (see *Figure 8*). Insert the 18-gauge needle attached to the large argon balloon in the rubber tubing between the flask and stopcock. The pipette bulb will inflate when the funnel and flask are full of argon. Remove the argon reservoir and return it to its rubber stopper. Remove the tubing from the flask and cover the side-arm with a pipette bulb. Turn the three-way stopcock to the open position and turn off the vacuum.

Before the next step, addition of the activated phosphoramidite, insert a small venting needle (1 inch long, 23-gauge) through the serum cap on the funnel to relieve the excess pressure caused by the injection of the activated phosphoramidite solution through the serum cap. If this needle is absent or is blocked, the excess pressure will cause some of the solution to be expelled through the sintered glass frit of the funnel. Activate the aliquot of phosphoramidite solution by tetrazole solution in an argon-filled mixing tube and then transfer to the support using the appropriate phosphoramidite syringe. Label the mixing tube with the number of the coupling and the phosphoramidite used. The set of empty tubes gives a check on the correctness of the sequence of coupling reactions.

Remember that the first deoxyribonucleoside is already attached to the support and that the first coupling reaction couples the second deoxyribonucleotide to the first. It is a good idea to have a check list at hand with the sequence written 3′ to 5′ (the direction of synthesis) so that reactions and washes can be checked off in the correct order as carried out (*Table 2*).

In preparation for the next coupling, when the syringes are returned to the desiccator after the latest coupling, the next phosphoramidite syringe should be placed adjacent to the tetrazole syringe at the front of the desiccator, and the next phosphoramidite solution to be used should be moved to the front of the rack next to the tetrazole solution.

Successful couplings depend on a fast accurate syringe technique. The activated phosphoramidite solution is very sensitive to atmospheric oxidation and moisture, and should be transferred to the support in less than 15 sec. Most people find these syringe operations unfamiliar and troublesome at first but they become second nature with a little practice. **Warning: the syringes should always be transported with needles downward to avoid puncturing colleagues, and to prevent the syringe plungers falling out of the barrels and breaking.** Syringes inserted into solutions under positive pressure (balloons) will fill with solution unless the plunger of the syringe is restrained. The rubber serum caps present some resistance to the needles. Aim for the centre of the serum cap. The top of the syringe and plunger should be securely held between the thumb and first two fingers (in a way similar to that of holding a pen). Guide the tip of the needle between the thumb and first finger of the other hand.

In this method of synthesis 30 μmol of activated deoxyribonucleoside-3′-O-phosphoramidite solution is required for efficient coupling to the 5′-hydroxyl groups of 1.5 μmol of support-bound deoxyribonucleoside. Consistently efficient couplings require consistently accurate dispensing of activated phosphoramidite solutions onto the support.

Table 2. Checklist for Monitoring Oligonucleotide Synthesis (N.B. The first nucleotide is already attached to the support: coupling 0)

Coupling	0	1	2	3	4	5	6	7	8	9	10	11	12	13	14	15	16	17	18	19	20	After the last coupling and oxidation
Sequence 3' → 5'																						
Capping: Ac$_2$O/lutidine 0.25 ml + DMAP soln 0.75 ml. React 2 min																						Do not cap
Wash MeCN 2 x 2 ml																						
Wash CH$_2$Cl$_2$ 2 x 1.5 ml																						
Funnel → trityl flask																						
Detritylation: DCA soln 2 x 1 ml																						Detritylate
20 sec treatment for 5'-A 2 x 1 ml																						Wash DCA
30 sec treatment for 5'-G,C,T																						
Wash DCA soln. 2 x 1/2 ml for 5'-A																						
4 x 1/2 ml for 5'-G, C and T																						
Funnel → organic waste flask																						
Wash CH$_2$Cl$_2$ or anhydrous																						Wash CH$_2$Cl$_2$
MeCN 5 x 2 ml																						Air dry
Funnel → coupling flask																						
Drying and introducing argon																						End of synthesis
Coupling: (a) 0.18 ml tetrazole soln (b) 0.15 ml phosphoramidite 3 min																						
Funnel → organic waste flask																						
Wash MeCN 2 ml																						
Oxidation: I$_2$ solution 1 ml 1 min																						
Wash MeCN 3 x 2 ml																						
Capping																						

Phosphite-triester Method

The simplest procedure is to weigh 30 μmol portions of phosphoramidites (~25 mg for A, G and C and ~22 mg for T) into labelled Eppendorf tubes, and then cover with small serum caps, vented with needles, evacuate and then fill with argon. When required for coupling, select the appropriate phosphoramidite and dissolve it in anhydrous acetonitrile (0.15 ml). Inject tetrazole solution (0.18 ml) and withdraw the contents of the tube. Inject it onto the support using the appropriate phosphoramidite syringe. This approach is adequate for the synthesis of short fragments such as hexa- or octanucleotide linkers, but the time-consuming weighings required for longer syntheses make this method too tedious for routine use.

A more practical approach is to make up phosphoramidite solutions of known molarity from which aliquots are removed and then activated with aliquots of tetrazole solution of known molarity. Diisopropyl phosphoramidite solutions in anhydrous acetonitrile are stable for at least 3 days when stored under positive argon pressure. On the day of synthesis dissolve phosphoramidites in anhydrous acetonitrile to give 0.2 M solutions and dissolve tetrazole to give a 0.5 M solution. Before tetrazole solution (0.18 ml, 90 μmol) is injected into the mixing tube it is essential to check that the venting needle just penetrates the serum cap and is not pushed deep within the tube, otherwise some tetrazole solution may be ejected *via* this needle when it is injected. Withdraw the tetrazole syringe and inject phosphoramidite solution (0.15 ml, 30 μmol) through the tetrazole solution, by placing the syringe needle at the base of the mixing tube. Rapidly withdraw the activated mixture and inject it through the vented serum cap of the funnel onto the support using the phosphoramidite syringe. The tetrazole and phosphoramidite syringes should be returned to the desiccator immediately after use.

The use of syringes to meter and deliver aliquots of phosphoramidite solution can present a potential problem. There are dead volumes of 50 – 70 μl in 1 ml syringes with 2 inch, 20-gauge needles and dead volumes of 20 – 30 μl in 0.25 ml syringes with similar needles. The syringes will over-deliver by this amount with each injection. Unless allowance is made for this, the stock solutions of phosphoramidites will be expended sooner than anticipated.

The simplest method is to withdraw correspondingly less solution into the syringe, but more consistent couplings are obtained by adding proportionately more phosphoramidite to each serum bottle.

Add an additional 30 μmol portion of phosphoramidite to serum bottles required for 2 – 4 couplings; add two additional portions to bottles required for 5 – 9 couplings and add three additional portions for bottles required for 10 – 14 couplings (*Table 3*).

It is convenient to weigh out the tetrazole and phosphoramidites the day before synthesis into dry serum bottles. These should be covered with medium sized serum caps, vented by small needles and evacuated, together with the mixing tubes, overnight in a desiccator. The next day bleed argon into the desiccator and inject sufficient anhydrous acetonitrile (0.15 ml for each 30 μmol amount) into each phosphoramidite bottle to give a 0.2 M solution. Once the phosphoramidites have dissolved replace the venting needles by needles attached

Table 3. The Tabulated Sequences of Three Syntheses and the Quantitites of Phosphoramidites Used.

3'								Coupling number								5'	Sequence	Orientation of funnel in stopper
	0	1	2	3	4	5	6	7	8	9	10	11	12	13	14	15		
	T	A	T	C	C	G	T	C	C	C	A	A	G	T	C	A	JB18	Right
	G	G	A	T	T	G	A	G	T	G	T	T	T	C	G	A	JB19	Front
	G	C	A	C	C	T	C	A	A	T	A	A	G	A	T	A	JB20	Left

60 mg of derivatised CPG support = 1.5 μmol of nucleoside

Number of 30 μmol phosphoramidite portions required		Number of 30 μmol phosphoramidite portions weighed	Amount weighed	Dissolved anhydrous acetonitrile
A	15	19 x 26 mg =	494 mg	19 x 0.15 ml = 2.85 ml
G	8	10 x 26 mg =	260 mg	10 x 0.15 ml = 1.5 ml
C	9	11 x 26 mg =	286 mg	11 x 0.15 ml = 1.65 ml
T	13	16 x 22 mg =	352 mg	16 x 0.15 ml = 2.41 ml
0.5 molar tetrazole		175 mg in 5 ml anhydrous acetonitrile per serum bottle		2 bottles used

Phosphite-triester Method

to argon balloons.

When 0.25 ml syringes are used, withdraw phosphoramidite solution to the 0.15 ml mark and tetrazole solution to the 0.18 ml mark. When 1 ml syringes are used add the same proportions of extra phosphoramidite, but withdraw phosphoramidite solution to the 0.1 ml mark and tetrazole solution to the 0.13 ml mark.

This compensation works well and there should be very little phosphoramidite solution remaining in the serum bottles when the synthesis has been completed. While the support is treated with activated phosphoramidite solution it is convenient to empty the contents of the dimethoxytrityl flask into a screw capped test-tube labelled with the number of the detritylation and the last phosphoramidite detritylated.

After 3 min transfer the funnel to the organic waste flask and return the coupling flask to the desiccator. Remove and discard the serum cap and venting needle from the funnel, remove the coupling mixture by filtration and wash the support with acetonitrile (2 ml).

Add the dark oxidising reagent (1 ml) and leave in contact with the support for 1 min. The capping mixture can be prepared during this time.

Remove the iodine solution and wash the support with acetonitrile (3 x 2 ml) until it and the effluent are colourless.

One cycle is now complete. The support is now ready for the next capping reaction.

The method described should give a cycle time of under 15 min. A 20-residue oligonucleotide can be assembled in about 5 h. In order to minimise errors, do not rush through a synthesis, work at a comfortable pace and interrupt the synthesis occasionally for short breaks (after every 6 or 8 couplings) to prevent the onset of tiredness.

The following procedures for the synthesis, deprotection and purification of oligodeoxyribonucleotides have been proven in our laboratory. They are simple, easy to follow and require minimal chemical skills. Successful synthesis should result provided that pure reagents are used, moisture is excluded from the coupling reaction and liquid transfers by syringe are fast and accurate.

5.5 Preparation for Synthesis

The following steps should be taken on the day before synthesis.
(i) Distil about 250 ml of h.p.l.c. grade acetonitrile from calcium hydride. Store in a dry stoppered flask sealed with Parafilm.
(ii) Prepare the mixing tubes from Eppendorf tubes, serum caps and needles as described.
(iii) Weigh out sufficient of each phosphoramidite and tetrazole into oven-dried serum bottles, cover with medium sized serum caps and vent with small needles.
(iv) Evacuate the mixing tubes and serum vials by pumping overnight in a desiccator.

On the day of synthesis carry out the following steps.
(a) Bleed argon into an evacuated desiccator containing mixing tubes, tetrazole and phosphoramidite bottles.
(b) Remove the syringes (six), filter funnel (one), filter flasks (three) and 30 ml serum bottles (two or more) from the oven and cool in another desiccator.
(c) Fill the reagent and solvent wash bottles. Stand them on paper towels.
(d) Clamp the organic waste and dimethoxytrityl collection flasks.
(e) Insert the funnel into a rubber stopper, place on the organic waste flasks.
(f) Add 1.5 μmol of deoxyribonucleoside-bound support.
(g) Inflate the small (six) and large (one) argon balloons as described.
(h) Fill two (or more if anhydrous acetonitrile washes are to replace the dichloromethane washes before the drying step) 30 ml serum bottles with anhydrous acetonitrile. Cover with a medium sized serum cap, place the needle of an argon-filled balloon through one of these serum caps.
(i) Dissolve the phosphoramidite in anhydrous acetonitrile to give 0.2 M solutions.
(j) Dissolve tetrazole in anhydrous acetonitrile to give a 0.5 M solution.
(k) Remove the venting needles from the serum caps covering the tetrazole and phosphoramidite solutions. Add an argon balloon to each bottle containing these solutions.
(l) Label and mark the tetrazole syringe at 0.18 ml and phosphoramidite syringes at 0.15 ml on 0.25 ml syringes, or at 0.13 ml for the tetrazole syringe and at 0.1 ml for phosphoramidite syringes when 1 ml syringes are used.
(m) Start synthesis.

5.6 Synthesis Cycle

Scale 1.5 μmol of support-bound deoxyribonucleoside.
(0) Place 60 mg of deoxyribonucleoside-derivatised support in a 2 ml sintered funnel of medium porosity[a].
(1) Capping: add 0.25 ml of Ac$_2$O/lutidine solution to the Eppendorf tube — dispense by Pipetman. Add DMAP solution to the 1 ml mark. Transfer to the support using a Pasteur pipette. React for 2 min. Filter (gentle vacuum).
(2) Wash the support with acetonitrile (wash bottle) 2 x 2 ml.
(3) Wash the support with dichloromethane (wash bottle) 2 x 1.5 ml.
(4) Transfer the funnel and vacuum lines to the dimethoxytrityl collection flask[b].
(5) Detritylation: check which deoxyribonucleotide is at the 5' terminus of the chain. (i) If 5'-A add 1 ml of DCA solution (wash bottle), react for 20 sec, filter, repeat. Wash the support with DCA solution (2 x 0.5 ml). (ii) If 5'-G, C or T add 1 ml DCA solution, react for 30 sec, repeat. Wash the support with 3 or 4 x 0.5 ml of DCA solution until the support and effluent are colourless.
(6) Transfer the funnel and vacuum line to the organic waste flask[b].

Phosphite-triester Method

(7) Wash the support with dichloromethane (wipe the nozzle of the wash bottle with a paper tissue before use) 5 x 2 ml, or wash the support with anhydrous acetonitrile (contained in 30 ml serum bottles) 5 x 2 ml.

(8) Drying the support: remove the funnel from the organic waste flask. Wipe the base of the funnel with paper tissue. Cover the funnel with a small serum cap. Place the funnel on the coupling flask. Cover the open limb of the three-way stopcock with a rubber pipette bulb. Apply **full vacuum**. Gently heat the support with a hairdryer or heat gun for 1 – 2 min.

(9) Introduction of an argon atmosphere: turn the three-way stopcock to isolate the flask and funnel from the vacuum. Insert the needle of the large argon balloon into the tubing between the flask and three-way stopcock. Remove the balloon and needle when the pipette bulb has inflated. Disconnect the rubber tubing, turn the stopcock to the open position and then turn the vacuum off. Cover the side arm of the coupling flask with a rubber pipette bulb. Insert the venting needle through the serum cap on the funnel[c].

(10) Coupling: withdraw an argon-filled Eppendorf tube from the desiccator, mark it with the coupling number and the deoxyribonucleotide to be coupled, check that the venting needle just penetrates the serum cap. Place it in a rack in front of the styrofoam rack. Select the appropriate phosphoramidite syringe and the tetrazole syringe. Withdraw tetrazole solution up to the mark on the syringe (0.18 ml on 0.25 ml syringes or 0.13 ml on 1 ml syringes), inject into the mixing tube and withdraw the syringe. Check the sequence to ensure that the correct phosphoramidite is about to be added with the correct syringe. Withdraw phosphoramidite solution to the mark on the syringe and inject the phosphoramidite solution through the tetrazole solution in the tube by placing the syringe needle at the bottom of the tube. Withdraw the contents of the tube in one transfer and inject onto the support. Allow the mixture to react for 3 min. Return both syringes to the desiccator immediately after use. It is convenient to empty the contents of the dimethoxytrityl flask into a labelled test-tube during this time. After 3 min replace the funnel on the organic waste flask and discard the serum cap and venting needle. Return the coupling flask to the desiccator. Remove the coupling mixture by filtration. Wash the support with acetonitrile (wash bottle) 1 x 2 ml.

(11) Oxidation: add iodine solution 1 ml (wash bottle) and react for 60 sec. Filter (the capping mixture can be prepared during this time). Wash the support with acetonitrile (wash bottle) 3 x 2 ml until the support and effluent are colourless.
[Return to capping (1).]

Notes:
(a) This quantity is for support with a loading of 25 μmol of deoxyribonucleoside per gram.
(b) These steps are omitted if dimethoxytrityl solutions are not to be collected.

(c) This is the safest place to interrupt a synthesis. If synthesis is interrupted, the support should be stored dry under argon in the serum-capped funnel placed on the argon-filled coupling tube in a desiccator at room temperature. The tetrazole and phosphoramidite solutions are stable and good for synthesis next day provided they are stored under argon balloons. The syringes should be washed well with acetone and returned to the oven. The mixing tubes should be evacuated by pumping overnight in a desiccator. When a synthesis is completed, residual phosphoramidite and tetrazole solution should be discarded with the organic waste. The syringes and serum bottles should be washed well with acetone and returned to the oven.

The three filter flasks should be rinsed with acetone and oven dried. The empty 30 ml serum bottles should be oven dried. The DMAP and acetic anhydride bottles should be tightly capped and sealed with Parafilm.

Unused reagents and solvents should be returned to the appropriate bottles. Any unused distilled acetonitrile (in 30 ml serum bottles) should be returned to the undistilled acetonitrile bottle. The support should remain in the funnel if the thiophenol treatment is to be used, if not, it should be transferred to an Eppendorf tube (1.5 ml).

Filter funnels can be re-used. They should be soaked in 5% sodium hydroxide solution for at least 24 h to destroy any oligonucleotides contaminating them, then washed well with water, then acetone and oven-dried. The funnels should be re-siliconised before use.

6. REMOVAL OF OLIGODEOXYRIBONUCLEOTIDE FROM SOLID SUPPORT, DEPROTECTION AND ISOLATION

When a synthesis has been completed, the support bears a fully protected oligodeoxyribonucleotide. The 5'-hydroxyl groups are dimethoxytrityl protected, the 3'-hydroxyl groups are linked to the support as succinate esters and the exocyclic amino groups on deoxyadenosine, deoxycytosine and deoxyguanosine residues are protected as amides.

The oligonucleotide has been assembled with a phosphotriester backbone, the internucleotide methyl esters on each phosphorus atom must be removed to generate the natural 3'-5' internucleotide phosphodiester linkage (between deoxyribose units (see *Figure 11*).

Deprotection is achieved in four stages.

(i) The dimethoxytrityl group is removed with DCA solution as the last step of the synthesis.
(ii) The internucleotide methyl esters are removed using an ammonium thiophenoxide treament (10).
(iii) An ammonia treatment at room temperature hydrolyses the 3'-succinate bond, this cleaves the oligonucleotide from the support and dissolves the released oligonucleotide.
(iv) The hot ammonia treatment hydrolyses the amides protecting the primary amino groups on deoxyadenosine, deoxycytosine and deoxyguanosine residues and also causes chain cleavage at sites of depurination (in a similar

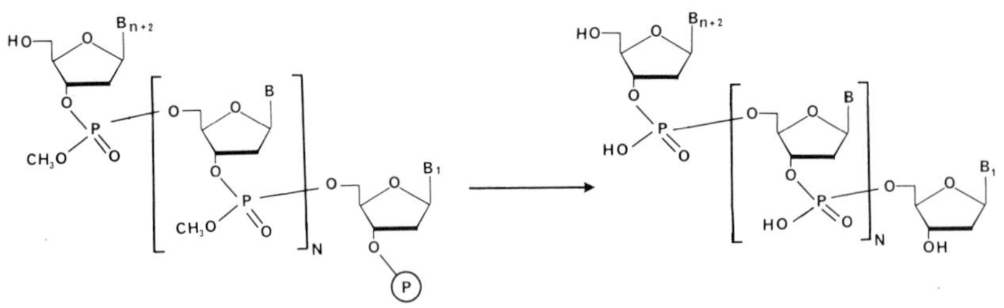

$B_{n+2} \ldots B_1$ = A^{Bz}, C^{Bz}, G^{iBu}, or T when a phosphotriester backbone is present

but = A, C, G, or T when a phosphodiester backbone is present in the deprotected deoxyoligonucleotide

Figure 11. Removal of oligodeoxyribonucleotide from solid support and deprotection.

fashion to piperidine cleavage in Maxam and Gilbert sequencing). Chain cleavage can reduce the yield of desired product but ensures that the oligonucleotide isolated is devoid of depurinated sites.

6.1 **Deprotection Procedure**

This method uses thiophenol to remove methyl ester groups.

(ii) Place the funnel containing the support-bound oligodeoxyribonucleotide on a filter funnel in a fumehood. Using a Pasteur pipette add 1 ml of thiophenol solution. **Warning: use gloves when handling this toxic reagent.** Cover the funnel with Parafilm. Gently agitate the support and let it stand for 1 h at room temperature. Remove the Parafilm from the funnel, and then remove the reagent by filtration. Destroy the filtrate by pouring it into a beaker containing bleach (10% sodium hypochlorite solution) then add the discarded pipette and Parafilm to the bleach. Wash the support with methanol (~30 ml) and then diethyl ether (~20 ml) and allow it to air dry.

(iii) Transfer the support to an Eppendorf tube (1.5 ml). Add 1 ml of concentrated ammonium hydroxide solution by Pasteur pipette. Cap the Eppendorf tube and seal with Parafilm. Return the ammonium hydroxide solution to the refrigerator. Allow the support to react with the ammonia solution for at least 1 h with periodic agitation. Remove the Parafilm. Centrifuge. Withdraw the ammonia solution of oligonucleotide and transfer to a glass 1 dram (4 ml; 40 x 13 mm) vial using a Pasteur pipette. Add 1 ml of fresh ammonia solution to the support. Vortex, centrifuge. Withdraw the solution and add it to that in the vial. Replace the screw cap on the vial, screw tight and seal well with Parafilm. Replace the ammonia solution in the refrigerator.

(iv) Place the vial in a small beaker (10 ml) a third full of hot water and place in a warm bath at 55°C overnight. Cool. **Warning: open carefully.** There should be a slight positive pressure. Distribute the solution equally between three Eppendorf tubes. Allow to stand open for at least 30 min and then evaporate to dryness using a Speed Vac Concentrator.

6.2 Alternative Deprotection Procedure

This procedure uses ammonia to remove methyl ester groups.

The thiophenol treatment can be avoided by allowing the support to be in contact with ammonia solution for at least 5 h at room temperature (or preferably overnight). This treatment cleaves the oligomer from the support and hydrolyses sufficient methyl esters to give an ammonia-soluble oligonucleotide. The oligonucleotide solution is transferred to a 1 dram (4 ml) glass vial and the support washed with another 1 ml of ammonia solution just as described in (iii). The hot ammonia treatment (iv) removes the amides and completes the removal of methyl esters.

The simpler procedure is perfectly adequate for deprotecting oligodeoxyribonucleotides containing up to 20 nucleotides, but it is advisable to use the thiophenol treatment on longer oligomers.

(v) Dissolve the contents of each tube in 300 µl sterile water. Vortex well. Centrifuge. Withdraw and combine the supernatant in an Eppendorf tube (1.5 ml). Discard the small amount of insoluble residue. Extract the supernatant with n-butanol (3 x 300 µl) to remove most of the benzamides. Evaporate the aqueous solution to dryness.

(vi) Ethanol precipitation: omit this step for oligodeoxyribonucleotides containing less than 10 nucleotides — go to (vii).

Dissolve the crude oligodeoxyribonucleotide in 100 µl of 0.5 M ammonium acetate solution with vortexing. Add 1 ml 95% ethanol. Vortex. Chill at −70°C for 30 min. Centrifuge at 4°C for 3 min. Remove the ethanol and wash the residue twice with cold (4°C) 95% ethanol. Under these conditions oligomers longer than decamers precipitate, benzamides and shorter truncated sequences remain in solution.

(vii) Dissolve the pellet in 200 µl of sterile water. Take a 2 µl aliquot and dilute to 1 ml with water and read the absorbance at 260 nm (A_{260}). This solution may be used to analyse the crude mixture by gel electrophoresis of a 5′ end-labelled sample. Evaporate the solution containing the bulk of the oligonucleotide to dryness.

(viii) Gel electrophoresis is used to separate the desired product from shorter fragments. Oligomers containing up to 20 nucleotides are purified on 20% polyacrylamide gels containing 7 M urea and run at constant voltage (1500 V) using 0.05 M Tris-borate buffer (pH 8.3) (Section 6.3). Oligomers containing 20–30 nucleotides should be run on 16% polyacrylamide gels and longer oligomers (30–50) on 12% gels. The mobilities of the oligonucleotides are determined by u.v. shadowing at 260 nm.

When synthesis is efficient and yields of crude oligonucleotide routinely exceed 100 A_{260} units, deprotection steps (vi) and (vii) may be omitted and a fraction (10–20%) of the material obtained at the end of step (v) may be purified by gel electrophoresis.

With efficient synthesis, it is unnecessary to deprotect all of the support-bound oligodeoxyribonucleotide. One third of the support can be deprotected and about a third of this purified by gel electrophoresis, the remaining support-bound protected oligodeoxyribonucleotide can be stored in Eppendorf tubes at $-18°C$. This may be deprotected later if more oligodeoxyribonucleotide is required or it may be extended by further synthesis. To resume synthesis allow the support to warm to room temperature, transfer to a dry siliconised funnel, repeat the support drying step and proceed with the next coupling reaction. Do not resume synthesis with a capping step because this will acetylate all the 5'-hydroxyl groups and render them inert towards further chain extension.

6.3 Purification of Oligonucleotides by Gel Electrophoresis

(i) Dissolve the crude oligonucleotide obtained from either step (vii) or step (v) of the deprotection procedure (Section 6.2) in 60–90 µl of sterile water.

(ii) Mix an aliquot of this solution (10 µl) with deionised formamide (20 µl) in an Eppendorf tube and denature by heating at 90°C for 3 min and then rapidly cool on ice before loading onto the gel.

(iii) Evaporate the remainder of the oligonucleotide solution to dryness and then store at $-18°C$.

(iv) Cast a 0.4 mm thick polyacrylamide gel (20%, 16%, or 12% depending on the length of the oligonucleotide) containing 7 M urea between siliconised glass plates with slots 10 mm wide.

(v) Load each slot with 10 µl of oligonucleotide solution using a Hamilton syringe. As a reference it is convenient to load about 3–4 µl of 90% formamide dye mixture in a spare slot.

(vi) Run the gel at constant voltage (1500 V) until the bromophenol blue is about two thirds down the gel (this is for 20% gels; the marker should be allowed to run further in gels containing less acrylamide).

(vii) Carefully transfer the gel to a sheet of Saran Wrap and visualise by u.v. shadowing. Place a t.l.c. plate which fluoresces at 254 nm beneath the gel. Short wave length u.v. illumination from above will cause the plate to fluoresce and regions in the gel containing oligonucleotides will absorb u.v. light and appear as dark bands against a fluorescent background.

(viii) The desired product should be the least mobile band and should be the major band present. Excise areas of the gel containing the oligonucleotide using a scalpel.

(ix) Place the three gel slices in an Eppendorf tube (1.5 ml), add elution buffer (1.5 ml) – 0.5 M ammonium acetate and 10 mM magnesium acetate in sterile water and incubate overnight at 37°C.

(x) Vortex the samples and then centrifuge for 5 min.

(xi) Withdraw the supernatant and transfer to an Eppendorf tube using a P1000 Pipetman.

(xii) Wash the gel slices with 0.5 ml of the elution buffer, vortex and then centrifuge.

(xiii) Remove the supernatant and combine it with the oligonucleotide solution.

(xiv) Filter to remove any gel fragments either through a small plug of siliconised glass wool forced into the end of a P1000 pipette tip or through a small Millipore disc (Millex HV or HV4). The Millipore filter discs are 0.45 μm pore size filter units which attach to syringes with Luer end fittings.
(xv) Load the gel elution solution into an inverted 3 ml disposable syringe (B-D 5571) using a P1000 Pipetman. The pipette tip should be inserted into the end of the syringe and the plunger slowly withdrawn as the contents of the pipette are discharged into the syringe.
(xvi) Attach the filter unit to the loaded syringe and then pump the solution through the filter and collect in Eppendorf tubes. Discard the syringe and filter unit after use.

The oligonucleotide can be isolated from the filtered solution either by a C_{18} SEP-PAK procedure or by ethanol precipitation of butanol concentrated elution solution. The SEP-PAK procedure is a much faster technique but the precipitation procedure requires less additional equipment.

6.3.1 *Isolation of Oligonucleotides*

(i) Reduce the volume of the elution solution to about 100 μl by repeated extraction with n-butanol.
(ii) Add 1 ml of 95% ethanol and chill the tube at $-70°C$ for 30 min, then centrifuge at 4°C for 10 min.
(iii) Withdraw the supernatant and wash the precipitate twice with cold 95% ethanol (0.5 ml at 0°C) centrifuge for 2 min, remove the ethanol and pump the residue to dryness.
(iv) Dissolve in 1 ml of sterile water, measure the absorbance at 260 nm, and evaporate the residue to dryness.

6.3.2 *Isolation using C_{18} SEP-PAK*

The C_{18} SEP-PAK cartridge consists of a small cylinder containing C_{18} reverse phase silica gel. Solutions are pumped through it by attaching a syringe with Luer end fittings to the longer tube end of the cartridge. It behaves as a small reversed phase chromatography column, when polar solvents are used the less polar components of a mixture are retained by absorption on the matrix while the more polar components are eluted.

The gel elution solution contains synthetic oligonucleotide in an aqueous electrolyte mixture. When this is passed through a prepared C_{18} cartridge the oligonucleotide is retained while the aqueous salts pass through it. The oligonucleotide is eluted by using a less polar solvent mixture after the column has been washed well with water.

Procedure. For each oligonucleotide to be purified follow these steps.
(i) Prepare the cartridge by passing h.p.l.c. grade acetonitrile (10 ml) through it followed by distilled or sterile water (10 ml).
(ii) Arrange seven Eppendorf tubes in a rack (to collect effluent washes).

Phosphite-triester Method

(iii) Prepare 1 ml portions of mixtures of methanol (h.p.l.c. grade) and sterile water (6:4) in the Eppendorf tubes.
(iv) Load the filtered gel elution solution into an inverted 3 ml disposable syringe using a P1000 Pipetman as described in the Millipore filtration step of the gel elution solution.
(v) Place the longer tube end of the SEP-PAK cartridge onto the inverted syringe and then pump the gel solution through it. Collect the effluent in approximately 1 ml aliquots in Eppendorf tubes (see step ii). Discard the syringe and pipette tip.
(vi) Using either a 5 ml or 10 ml disposable syringe, wash the cartridge with distilled or sterile water (5 ml). Collect the effluent in 1 ml portions in Eppendorf tubes.
(vii) Load 1 ml of methanol/water (6:4) into a new 3 ml disposable syringe using a P1000 Pipetman. Attach the C_{18} cartridge, pump the solution through the cartridge and collect the eluant in the empty tube that contained the methanol/water mixture.
(viii) Repeat step (vii) for the two remaining methanol/water portions then discard the SEP-PAK and syringe.
(ix) Set the u.v. spectrophotometer to measure absorbance at 260 nm.
(x) Fill two cuvettes with 1 ml methanol/water and adjust to zero.
(xi) Empty the sample cuvette and read the A_{260} of the three methanol/water fractions, and then that of the five water wash fractions, and then that of the two fractions of gel elution solution that passed through the cartridge. At least 80–90% of the oligonucleotide should elute in the first methanol/water fraction, with very little in the second and none in the third. The other solutions should have zero absorbance. Evaporate the solutions containing oligonucleotide to dryness using a Savant Speed Vac Concentrator.

7. SCALE UP OR DOWN, MULTIPLE PARALLEL SYNTHESIS AND SYNTHESIS OF MIXED SEQUENCE OLIGODEOXYRIBONUCLEOTIDES

7.1 Scale Up or Down

Efficient synthesis of fragments containing 15–30 nucleotides using 1–2 μmol of support-bound deoxyribonucleoside should yield about 100–200 A_{260} units of crude oligodeoxyribonucleotide. Purification, by gel electrophoresis, of approximately 10–20% of the total crude mixture should yield at least 0.5–2.0 A_{260} units of pure oligonucleotide. This is more than adequate for most purposes such as making primers for sequencing, for mutagenesis or for gene synthesis (see Chapter 1).

If larger quantities of pure oligodeoxyribonucleotide are required (e.g., for physical studies) it is easy to scale up the described method. Larger funnels (15 ml capacity), more support and proportionately more reagents should be used and an even larger amount of solvent should be used during the washing steps. The solvent wash volumes in 15 ml funnels should be at least 3–6 times larger than those used in 2 ml funnels to ensure efficient washing of the larger sintered frit and the greater surface area of the funnel's walls. Synthesis in excess of 5 μmol of bound deoxyribonucleoside can become expensive as more than 100 μmol of phosphoramidite is consumed in each coupling reaction.

Scaling down to the 1.0 μmol level using controlled pore glass supports in 2 ml funnels is simple. Use proportionately less support and reagents but maintain the volumes of solvent washes as described for the 1.5 μmol scale.

Synthesis on a smaller scale becomes more difficult as the quantitites of tetrazole and phosphoramidite solutions that have to be dispensed and transferred by syringe approach the dead volumes of the syringe and needle. This can lead to errors in syringe delivery and cause poor couplings.

Using more dilute solutions of phosphoramidite (0.1 M) and tetrazole (0.25 M) and allowing longer for the coupling reaction (5 min) is one way of reducing the problem of accurately dispensing small volumes by syringe. The 1.5 μmol scale is convenient for most syntheses. It is inadvisable to consider reducing the scale of a synthesis unless the method is routinely working well (>96% coupling efficiencies at each cycle) and the desired product is the major band on the preparative gel.

It is worthwhile considering that in a synthesis with an average coupling efficiency of 95%, after seven couplings the number of 5'-hydroxyl groups available for coupling will have been reduced by a third. So after 'n' couplings starting with 1.0 μmol of derivatised support, there will be approximately the same number of hydroxyl groups as in a similar synthesis starting with 1.5 μmol of derivatised support after (n + 7) couplings.

Therefore small-scale synthesis should be used for shorter fragments of 12 nucleotides or less while synthesis involving 30 or more condensations should start with 2 or 2.5 μmol of derivatised support to give a reasonable yield of product.

7.2 Multiple Simultaneous Synthesis

The time required for the deprotection and purification is the rate-determining step in obtaining oligodeoxyribonucleotides since solid-phase synthesis is so rapid. Considerable time can be saved by synthesising, deprotecting and purifying oligodeoxyribonucleotides in batches as opposed to one at a time. The simplest method is to connect up to six filter flasks in series to a common vacuum line but this becomes more cumbersome since a lot of space is required if more than two filter flasks are used. A more compact approach involves using more than one 2 ml filter funnel per filter flask by employing 'Twistfit' stoppers with two or three holes. Two funnels can be accommodated in a size 5 stopper seated in a 125 ml flask and three funnels in a size 7 stopper seated in a 500 ml flask. In this way 2, 3, 4 or 6 funnels can be simultaneously manipulated.

Controlled pore glass supports should be used with 2 ml funnels. It is important that the funnels drain at similar rates otherwise the more porous funnels will pull air through the support (this can cause poor couplings on humid days) while the less porous ones are still filtering. Funnels that filter more slowly will cause the support to be in contact with reagent solutions for longer than the times specified. It is easy to measure before hand the time each funnel takes to empty and then match funnels with similar drainage rates for batch synthesis.

Reagents and solvents are added to the funnels in rapid succession, are allowed to remain in contact with the support for specific times and are then removed simultaneously by filtration.

This method has two minor disadvantages, the more funnels in use the more the timing of the reactions becomes progressively less precise. Also individual dimethoxytrityl solutions cannot be collected for assay.

An additional 20–30 sec exposure of the support-bound oligodeoxyribonucleotide to the coupling, oxidising and capping reagents will not affect the synthesis adversely. However, some 5' N-benzoyldeoxyadenosine residues exposed to an additional 20–30 sec of dichloroacetic acid solution may depurinate resulting in a reduced yield of oligodeoxyribonucleotide. The inability to collect individual trityl solutions is not important, since the success of each coupling can be determined by inspection. Furthermore, it is inadvisable to attempt multiple synthesis until single oligodeoxyribonucleotide synthesis is working satisfactorily. Then the collection and assay of dimethoxytrityl solutions becomes redundant. Successful synthesis depends on good organisation to avoid confusion and error.

The funnels should be clearly labelled and orientated so that the same funnel is always closest to the operator. Tabulate the sequences so that the phosphoramidites required for the next round of coupling are clearly visible at a glance (*Table 3*).

If three funnels are in use, select three mixing tubes, label each with the sequence number, the phosphoramidite about to be coupled, and the number of the coupling.

Inject an aliquot of tetrazole solution into each tube and then inject an aliquot of phosphoramidite solution into the first tube, rapidly withdraw the activated coupling mixture and inject it into the first funnel. Then activate and add the second phosphoramidite and then the third. Then start timing the coupling reaction.

Each additional funnel adds approximately 4 min to the reaction cycle. One funnel should give a cycle time of about 14 min, two of 18 min and three of about 22 min. It is quite feasible in a day to synthesise one 30-mer, or two 20-mers, or three 16-mers, or four 12-mers.

Synthetic oligodeoxyribonucleotides for use as mutagenic primers are designed to be self-complementary to the site of interest except for the mismatch that represents the mutation (see Chapter 1). The oligonucleotide is designed so that the mismatch is located near the middle of the molecule. For a given experiment often more than one mutagenic primer is required and a family of similar oligonucleotides differing in sequence near the middle of the molecules can be synthesised using batch methods.

Table 4 illustrates three mutagenic nonadecanucleotide primers differing only in sequence at the ninth and tenth nucleotides (from the 3' end). These can be synthesised individually in three 2 ml funnels using 1.5 μmol of deoxyguanosine-derivatised controlled pore glass per funnel. A considerably faster method is to perform the first seven condensations in a 15 ml funnel using 4.5 μmol of derivatised support and then divide the dry support equally between three 2 ml

Table 4. Three Mutagenic Nonadecanucleotide Primers.

3'		Coupling number		5'
0	8	9	18	
G	G	T	T	AB1
G	T	T	T	AB2
G	G	A	T	AB3

funnels before the eighth round of couplings and then use the batch method until the end of the synthesis.

For the first seven couplings, 90 μmol of phosphoramidite solution (0.45 ml) is activated with tetrazole solution (0.54 ml). The activated solution is transferred to the support using a 1.0 ml syringe. The tetrazole syringe is marked at 0.49 ml and the phosphoramidite syringe is marked at 0.4 ml to deliver the required volumes.

After coupling number seven 0.25 ml syringes marked at 0.15 ml and 0.18 ml or 1.0 ml syringes marked at 0.1 ml and 0.13 ml are used for phosphoramidite and tetrazole solutions respectively.

When the support has been divided, the drying step should be repeated before argon and the activated phosphoramidite solution are introduced.

7.3 Synthesis of Mixed Sequence Probes

Synthetic oligonucleotides can be used as probes to isolate and clone specific eukaryotic genes for which the amino acid sequence of their protein product is known. In most cases, the degeneracy of the genetic code will not allow the unambiguous prediction of an oligonucleotide sequence from the corresponding amino acid sequence. This problem can be solved by the use of a mixed sequence, synthetic oligonucleotide probe (see Chapter 1). In such a probe, degenerate positions can be occupied by a mixture of 2, 3 or 4 nucleotides, so that the final population of molecules is a mixture of related but non-identical oligonucleotides. The synthesis of such a probe can be accomplished by using a mixture of the appropriate deoxyribonucleoside phosphoramidites in the coupling reaction when degenerate sites are required.

In practical terms sufficient equimolecular amounts of C/T, A/G, A, G, C, T or any other mixture are accurately weighed into 5 ml serum bottles and then dissolved to give 0.2 M solutions in anhydrous acetonitrile. An extra syringe should be dedicated to each extra serum bottle. The solutions are stored under argon balloons and aliquots are withdrawn and activated as described for single phosphoramidite solutions.

7.4 Nucleoside-specific Coloured Trityl Cations

The colour of acidic solutions containing triaryl methyl cations is influenced by the nature of the aryl substituents attached to the central carbon atom. Fisher and Caruthers have demonstrated that almost any colour can be produced by subtle substitution (11).

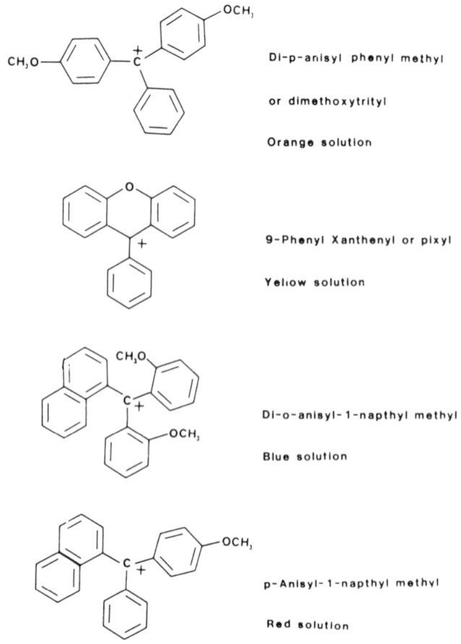

Phosphite-triester Method

Di-p-anisyl phenyl methyl or dimethoxytrityl
Orange solution

9-Phenyl Xanthenyl or pixyl
Yellow solution

Di-o-anisyl-1-napthyl methyl
Blue solution

p-Anisyl-1-napthyl methyl
Red solution

Figure 12. Coloured triaryl methyl cations.

If a different triaryl methyl group is used to protect the 5'-hydroxyl groups of the four deoxyribonucleotide monomers, then a collection of consecutive coloured 'trityl' solutions would give a simple, rapid check on the veracity of the sequence of synthetic oligodeoxyribonucleotides. Moreover, because, the large difference in visible absorption spectra for the various triaryl methyl cations, these solutions can be used to monitor the efficiency of each coupling reaction and also the relative efficiencies when more than one nucleotide has been coupled in the synthesis of probes of mixed sequences.

Figure 12 illustrates four promising candidates for 5'-protection of the monomers. Their 5'-ethers are cleaved at similar rates when solutions of trichloro- or dichloroacetic acid are used, and the solutions produced display a good range of colours.

However, it remains to be seen whether these or similar derivatives become widely accepted and commercially avaiable, or whether a colour standard is agreed upon (i.e., yellow for A, orange for G, blue for C and red for T). (Coloured deoxynucleoside monomers are now commercially available from Cruachem.)

7.5 Variations in Procedures

Oligonucleotide synthesis has evolved rapidly in the last 3 years. Improved supports, and more efficient and more selective reagents have been introduced to good effect. The synthesis of oligodeoxyribonucleotides containing more than 40 nucleotides by monomer additions is now possible by manual or automated methods.

The risk of some depurination, occurring when N-benzoylated deoxyadenosine residues are detritylated, is one of the last remaining problems of oligodeoxyribonucleotide synthesis. Caruthers introduced zinc bromide in nitromethane solutions as a safe, aprotic acid, detritylating agent the use of which leads to negligible depurination. Unfortunately this has proved to be a slow reagent for oligonucleotides attached to silica supports. Detritylation is particularly slow with deoxycytidine residues — as long as 20 – 30 min, which is much longer than the time required to carry out the rest of one cycle of nucleotide addition using phosphite methods.

Trichloroacetic acid (TCA) in dichloromethane, or nitromethane has been used for a few years. Detritylation is rapid, usually within 60 – 90 sec, the purines detritylate faster than the pyrimidines.

DCA in dichloromethane is currently the most popular detritylating agent. It is not such a strong acid as TCA but detritylation times are similar to those of TCA, and it appears to cause less depurination. Comparable synthesis gives up to 50% higher yields of crude oligonucleotide and much less shorter fragments are observed in preparative gels when DCA is used instead of TCA.

Matteucci (12) has recently introduced sterically hindered N^6-dialkyl aminomethylene derivatives, particularly N^6-di-n-butylaminomethylene, as a depurination-resistant N^6-protecting group for deoxyadenosine residues (see Chapter 4 for preparation). This promising novel protecting group is stable to the reaction conditions of phosphoramidite synthesis and is efficiently removed by hot ammonia.

3′-N-benzoyldeoxyadenosine residues are much less susceptible to depurination than are the 5′ residues. Many oligonucleotides of reasonable length-20 – 30 nucleotides, have been synthesised using N-benzoyldeoxyadenosine-derivatised supports using TCA as the detritylating agent. Recently two 43 long oligonucleotides were synthesised in our laboratory using N-benzoyldeoxyadenosine-derivatised supports and DCA. However the depurination problem should be eliminated if N^6-dialkylaminomethylene-protected deoxyadenosine supports become commercially available.

Synthesis using a 20-fold excess of activated phosphoramidite solution over support-bound deoxyribonucleoside may seem excessive. It is better to carry out synthesis on a small scale with a reasonable excess of coupling reagent and obtain consistently high coupling yields, than to carry out the synthesis on a larger scale using less coupling reagent, and obtain variable yields and a more difficult purification at the end of the synthesis. The 20-fold excess of activated phosphoramidite builds a margin of safety into the system. Small errors in syringe delivery, loss of some coupling reagent by dripping through the filter and atmospheric humidity are tolerated with little loss in coupling efficiency.

Consistently good couplings can be achieved using a 10-fold excess of activated phosphoramidite with controlled pore glass supports if a totally enclosed reaction vessel is used. Such is the case in column-based synthesis systems including commercial DNA synthesisers.

With sufficient care consistently good couplings can be obtained by experienced operators using a 15-fold excess of coupling mixture, but it is inadvisable to use less excess with the funnel method of synthesis.

8. CHEMICALS AND EQUIPMENT

8.1 Oligodeoxyribonucleotide Synthesis Equipment

Filter flasks, pyrex	125 ml	
	500 ml	
Filter funnels, pyrex medium porosity	2 ml	
	15 ml	
Rubber stoppers	size 5	Twistfit
	size 7	Twistfit
Three-way stopcock	4 mm bore	Nalge
Serum bottles dark glass	5 ml	Aldrich
clear glass	30 ml	Aldrich or Pierce
Rubber serum caps small	7 mm size	
medium	13 mm size	Aldrich or Sigma
large	24 mm size	
Glass tuberculin syringes	1 ml	Becton-Dickinson
	or 0.25 ml	
Stainless steel syringe needles 20-gauge 2 inch long Luer-lok hub		Becton-Dickinson
Disposable plastic syringes	3 ml	Becton-Dickinson
	5 ml or 10 ml	
Disposable syringe needles	22-gauge 1 inch	Becton-Dickinson
	18-gauge 1 inch	Becton-Dickinson
Wash bottles	60 ml	Nalge
	or 125 ml	

Pasteur pipettes
Rubber pipette bulbs
Rubber balloons
1 cylinder of oxygen-free argon and regulator
1.5 ml Ependorf tubes
Rack for Eppendorf tubes
Desiccator
Vacuum pump
Vortex mixer

8.2 Oligodeoxyribonucleotide Deprotection

1 dram (4 ml; 40 x 13 mm) vials screw cap	Kimble
Savant Speed Vac evaporator	Savant Industries
Waterbath 55–60°C	

8.3 Oligodeoxyribonucleotide Purification by Gel Electrophoresis

High voltage power pack
20 x 40 cm gel kit with 0.4 mm spacers and
   1 cm slot formers
Yellow 3M tape for making gel sandwich

U.v. lamp 254 nm (in dark room)
Kieselgel F254 t.l.c. plates   Merck Art 5554
Saran wrap
C$_{18}$ SEP-PAK cartridges   Waters
U.v. spectrophotometer to read A_{260}

8.4 Reagents for Oligodeoxyribonucleotide Synthesis

	USA	Europe
Deoxyribonucleoside-3'-O-diisopropyl-phosphoramidites	Applied Biosystems American Bio Nuclear	Applied Biosystems Cruachem
Deoxyribonucleoside-derivatised supports		
Long chain alkylamine, controlled pore glass	Applied Biosystems American Bio Nuclear Cruachem	Applied Biosystems Cruachem
Fractosil	Applied Biosystems American Bio Nuclear Cruachem Biosearch	Applied Biosystems Cruachem
Dichloroacetic acid	Aldrich BDH	BDH
Acetic anhydride	Aldrich	
2,6-Lutidine	Aldrich Cruachem	Cruachem
4-Dimethylaminopyridine	Aldrich	
Sublimed tetrazole	Aldrich	
Sublimed iodine	Aldrich Fisher	
Concentrated ammonia solution	Baker	
Thiophenol	Aldrich	
Triethylamine	Baker, Fisher, Aldrich	
Dioxane	Baker, Fisher, Aldrich	

8.5 Solvents for Oligodeoxyribonucleotide Synthesis

H.p.l.c. grade acetonitrile (some distilled from calcium hydride)	Baker, Fisher
H.p.l.c. grade THF tetrahydrofuran (some distilled from calcium hydride)	Baker, Fisher
H.p.l.c. grade dichloromethane	BDH, Baker Fisher
H.p.l.c. methanol	Baker
Anhydrous diethylether	

Phosphite-triester Method

8.6 Reagents for Preparing Deoxyribonucleoside-3'-O-phosphoramidites

Methanol	Baker
Phosphorus trichloride	BDH, Aldrich
Methyl phosphodichloridite (MeOPCl$_2$)	Aldrich
Diisopropylamine	Aldrich
Dichloromethane	
Phosphorus pentoxide	Aldrich, Baker
Alumina	
Chloro-N,N-diisopropylaminomethoxyphosphine	American Bio Nuclear, Aldrich
5'-O-dimethoxytrityl-2'-deoxyribonucleosides	Pharmacia PL, Cruachem, Vega Biosearch
Diisopropylethylamine	Aldrich
Ethylacetate	Baker
Triethylamine	Baker
Kieselgel 60	Merck Art 7734
Sodium chloride	
Sodium carbonate	
Anhydrous sodium sulphate	

8.7 Reagents for Preparing Deoxyribonucleoside-derivatised Supports

Long chain alkylamine derivatised, controlled pore glass support	Pierce
Fractosil 500	Merck
3-Aminopropyltriethoxysilane	Aldrich
Ninhydrin	Aldrich
n-Butanol	Aldrich
5'-O-dimethoxytrityl-2'-deoxyribonucleoside	Pharmacia PL, Vega, Cruachem, Biosearch
Succinic anhydride	Aldrich
Anhydrous pyridine	Aldrich
Citric acid	Aldrich
Toluene	Aldrich
Hexanes	Aldrich
5'-O-dimethoxytrityl-2'-deoxynucleoside-3'-O-succinate	Cruachem
p-Nitrophenol	Aldrich
Dicyclohexylcarbodiimide	
N,N-dimethylformamide	
Triethylamine	

9. REFERENCES

1. Letsinger,R.L., Finnan,J.L., Heavner,G.A. and Lunsford,W.B. (1975) *J. Am. Chem. Soc.*, **97**, 3278.
2. Matteucci,M.D. and Caruthers,M.H. (1981) *J. Am. Chem. Soc.*, **103**, 3185.
3. Beaucage,S.L. and Caruthers,M.H. (1981) *Tetrahedron Lett.*, **22**, 1859.
4. Adams,S.P., Kavka,K.S., Wykes,E.J., Holder,S.B. and Galluppi,G.R. (1983) *J. Am. Chem. Soc.*, **105**, 661.
5. McBride,L.J. and Caruthers,M.H (1983) *Tetrahedron Lett.*, **24**, 245.
6. Martin,D.R. and Pizzolato,P.J. (1950) *J. Am. Chem. Soc.*, **72**, 4585.
7. Dörper,T. and Winnacker,E.-L. (1983) *Nucleic Acids Res.*, **11**, 2575.
8. Tanaka,T. and Letsinger,R.L. (1982) *Nucleic Acids Res.*, **10**, 3249.
9. Hofle,V.G., Steglich,W. and Vorbruggen,H. (1978) *Angew. Chem.*, **90**, 602.
10. Daub,G.W. and van Tamelen,E.E. (1977) *J. Am. Chem. Soc.*, **99**, 3526.
11. Fisher,E.F. and Caruthers,M.H. (1983) *Nucleic Acids Res.*, **11**, 1589.
12. Froehler,B.C. and Matteucci,M.D. (1983) *Nucleic Acids Res.*, **11**, 8031.

UPDATE

Recently, a novel form of deoxyribonucleoside-3'-O-phosphoramidite has become commercially available. These are the β-cyanoethyl-N,N-diisopropyl phosphoramidites (13), in which a β-cyanoethyl protecting group replaces the standard methyl protecting group for the internucleotide phosphate. This eliminates the need to use thiophenol for the deprotection of methyl esters at the end of a synthesis.

Instead, oligonucleotides prepared using β-cyanoethyl phosphoramidites can be fully deprotected using only concentrated ammonium hydroxide solution. A 1 hour treatment at room temperature cleaves the succinate linkage and also removes all the phosphate protecting groups by means of a reaction involving a β elimination mechanism. Treatment at 55°C overnight then completes the deprotection by removal of the base protecting groups. In our experience, β-cyanoethyl phosphoramidites display comparable stability, both as dry powders and in anhydrous acetonitrile solution, to that of the corresponding methyl ester-protected phosphoramidites, and they are just as efficient in the internucleotide coupling reactions. β-Cyanoethyl phosphoramidites are now available from American BioNuclear, Biosearch, Biosyntec, Cruachem and Vega Biochemicals.

We have also recently found that the time of reaction for internucleotide couplings can be reduced from 3 minutes to 1 minute without detriment to the synthesis, using either type of phosphoramidite.

Aldrich now supplies anhydrous acetonitrile in 100 ml and 800 ml amounts (catalogue number 27, 100−4) with a water content of less than 0.005%. This acetonitrile can be used directly, without a prior distillation step, to dissolve tetrazole and phosphoramidites. Anhydrous acetonitrile is very hygroscopic. Therefore, a bottle should be opened for the shortest possible time and, after use, all the air inside the bottle should be promptly replaced with argon (using a balloon and needle) before closing and sealing the bottle with Parafilm. A bottle that has been opened and purged with argon more than four times should no longer be considered to be anhydrous. It may therefore be more economical to use 100 ml bottles of anhydrous acetonitrile for oligonucleotide synthesis.

Finally, oligonucleotides can be eluted from C_{18} SEP PAK cartridges (see section 6.3.2) using a mixture of acetonitrile (h.p.l.c. grade) and sterile water (1:4).

13. Sinha,N.D., Biernat,J., McManus,J. and Köster,H. (1984) *Nucleic Acids Res.*, **12**, 4539.

CHAPTER 4

Solid-phase Synthesis of Oligodeoxyribonucleotides by the Phosphotriester Method

BRIAN S. SPROAT and MICHAEL J. GAIT

1. INTRODUCTION

The beauty of solid-phase synthesis of DNA fragments lies in the fact that, by simple washing, excess reagents are easily removed from the reaction product which is attached to an insoluble support as a growing chain. However, since all purification steps are left to the very end of the assembly, this places certain demands on the method in general. Thus, protecting groups must be stable to the various conditions used in the assembly cycles, but must also be cleanly and quantitatively removed at the appropriate time. Internucleotide coupling reactions should be as efficient and as selective as possible to prevent the accumulation of failure sequences, truncated sequences and other impurities, which would lead ultimately to a poor yield of the desired product.

The aim of this chapter is to enable a non-specialist to assemble oligodeoxyribonucleotides by the phosphotriester method. Protocols for the preparation of the various starting materials are included for the more ambitious. This is often the best way of ensuring the highest purity of starting materials and hence obtaining routinely reliable oligodeoxyribonucleotide syntheses; the methods involve no intrinsically dangerous procedures. Sources of commercially available starting materials are included for those who do not wish to prepare their own. It is the authors' experience, however, that the quality of such materials (particularly deoxyribonucleotide monomers and solvents) varies widely from supplier to supplier and occasionally from batch to batch. It is very important that the non-specialist be aware of possible pitfalls and, wherever possible, these are highlighted (see also Chapter 1). The advantage of the particular method of DNA synthesis described here is that quite long fragments can be assembled in a relatively short time (~ 8 h for a 20-residue oligonucleotide) using relatively inexpensive equipment and with a minimum number of solvents and reagents. The method has proved routinely useful in a large number of laboratories and is extremely versatile. It is easily adapted for preparing quantities of oligodeoxyribonucleotides from 20 μg to 20 mg or more.

2. GENERAL METHODS AND PRECAUTIONS
2.1 Sources of Potential Trouble
The experimental procedures described here are those in current use in the

authors' laboratory. They have been developed to minimise the assembly time for oligodeoxyribonucleotides and maximise their yield by employing the best protecting groups available and using a minimum of solvents, reagents and equipment. Since trouble-shooting in solid-phase synthesis is particularly difficult, procedures should be rigorously adhered to and any changes made one at a time. In the event of a failed synthesis some possible causes of trouble can then be eliminated immediately (see also Chapter 1, Section 5).

Variability in commercial solvents, analytical grades included, where the presence of certain impurities can wreak havoc with an assembly, can be largely suppressed by careful distillation and checking. Prime culprits are primary and secondary amine contaminants in N,N-dimethylformamide and pyridine, which can cause undesired cleavage of the succinate linkage between the growing DNA chain and the support. The linkage must be preserved during chain assembly and cleaved only at the end of the synthesis by use of a standard oximate solution ($t_{1/2} \sim 1$ min). The problem of unwanted linkage cleavage is particularly serious when carrying out small-scale synthesis (<1 μmol of support-bound deoxyribonucleoside), which is often favoured because of reduced cost or because only small quantitites of the final product are required. The presence or absence of amines in solvents can be assayed by the ninhydrin test (Appendix I). Of equal importance is the purity of the monomers, since they are usually used in substantial excess over the support-bound species that is to be phosphorylated. The presence of even a small quantity of a highly reactive impurity can cause a considerable side reaction (e.g., base-modification, truncation). Such reactions repeated over a few cycles will lead to a rapid attenuation of the amount of the desired sequence.

Molecular sieves, often used for keeping solvents dry, should be avoided at all costs since prolonged contact with pyridine and other basic solvents such as N,N-dimethylformamide (DMF) can lead to metal ions (notably Fe^{3+}) being taken into solution. Metal ions cause a number of uncharacterised but severe side-reactions which may include loss of 5'-protecting groups, cleavage of phosphate protecting groups and modification of the heterocyclic bases followed by internucleotidic cleavage during final deprotection. It is sufficient for such solvents to be stored without additives in sealed containers, preferably under nitrogen or argon.

Purity of solvents and reagents is of the utmost importance as far as reliability and reproduciblity of the method are concerned.

2.2 The Support

Numerous supports are available for solid-phase synthesis of DNA by the phosphotriester method; among them are the gel resins of polystyrene (1) and polacryloylmorpholide (2), a composite support of polydimethylacrylamide and Kieselguhr (3–6), porous, inert, inorganic materials such as silica (7,8) and controlled pore glass (9–11) and finally, for use in rather special circumstances, cellulose paper (12,13).

Solid-phase synthesis involves repeated cycles of addition of a solvent or reagent to a solid support followed by removal of the solvent or reagent by washing. The most efficient way to wash a support is to pack it into a column and flow solvents and reagents through it. The highly swollen gel resins are not ideally suited for such a system. Thus we use materials that do not pack down under pressure of solvent and also avoid fine mesh materials, which can cause clogging of frits and generate high back-pressures. The favoured supports for column use in our experience are long chain alkylamine controlled pore glass (LCAA/CPG), polydimethylacrylamide-Kieselguhr, and Whatman 3MM chromatography paper, each of which is suited to a particular need.

Long chain alkylamine controlled pore glass (Pierce, pore diameter 500 Å, particle size 125–177 μm) is arguably the best support for the preparation of small quantities of oligodeoxyribonucleotides (50 μg – 2 mg, ~1 – 50 A_{260} units). The support can be readily functionalised at a level of about 10–30 μmol of deoxyribonucleoside per gram of support (*Figure 1*). It is an ideal support in that it is chemically inert, non-swellable, and relatively non-polar. In our laboratory a purine-rich 37-residue oligonucleotide has been prepared in good yield (~2%) on this support and a 51-residue oligonucleotide has been prepared by others (14) using the phosphite-triester approach. The fully functionalised supports are commercially available (e.g., from Cruachem) or can be prepared according to the procedure given in Section 3.1.2.

The polydimethylacrylamide-Kieselguhr support developed in this laboratory by the group of R.C. Sheppard and co-workers (15) is best reserved for larger scale preparations (~5–50 mg of product) where the product is required for structural studies (X-ray crystallography and n.m.r. spectroscopy, for example). There are two supports available commercially, one of which can be functionalised to a level of about 90 μmol of deoxyribonucleoside per gram of support and a 'double loading' version giving a loading of about 180 μmol per gram. The linkage used with this support is illustrated in *Figure 2*. The support consists of fabricated beads of Kieselguhr containing very large pores (several thousand Ångstroms in diameter) in which cross-linked polydimethylacrylamide resin (15) has been prepared. The support contains a controlled amount of ester functionalities introduced in the polymerisation reactions used in its preparation as acryloyl sarcosine methyl ester. Derivatisation (see Section 3.1.1) involves an initial treatment with ethylene diamine followed by addition of two glycine spacers and finally addition of a protected 2'-deoxyribonucleoside-3'-O-succinate. Un-

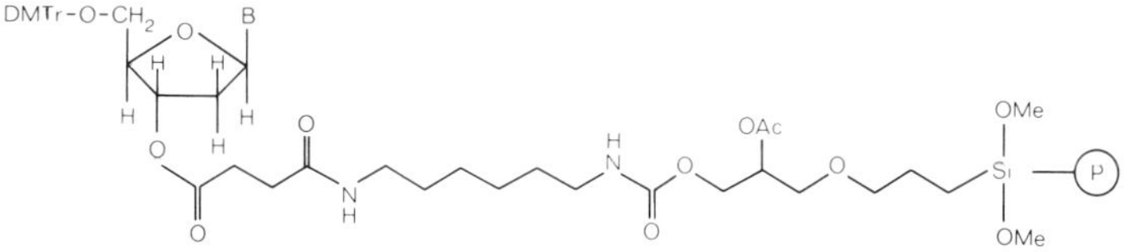

Figure 1. The standard linkage of deoxyribonucleoside to long chain alkylamine controlled pore glass (LCAA/CPG). A newer version of LCAA/CPG has a second hexamethylene diamine group linked as a urethane (-NH-C(=O)-O-) in place of the secondary acetate group.

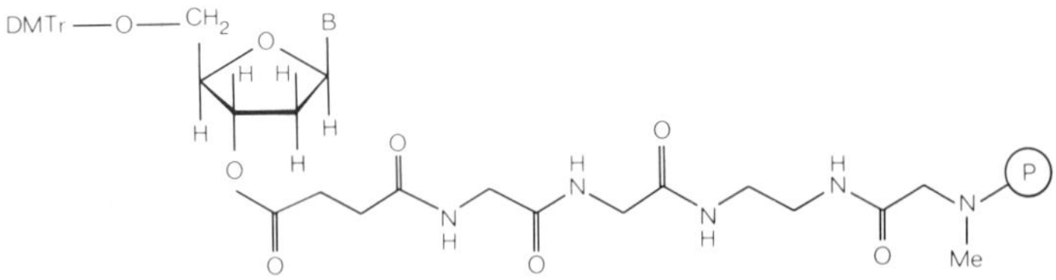

Phosphotriester Method

Figure 2. The gly-gly-ethylene diamine spacer used in the linkage of a deoxyribonucleoside to polydimethylacrylamide-Kieselguhr.

functionalised support is manufactured by Victor Wolf, Ltd. (Clayton, Manchester, UK) as Wolfchem SR 108 (normal) and Wolfchem SR 109 (double-loaded) and is available from several suppliers. Fully and partially functionalised material is also available commercially.

The use of Whatman 3MM paper discs (13) enables very small quantities of oligodeoxyribonucleotides to be prepared ($\sim 0.5 - 2.0$ A_{260} units). This support is ideally suited to the simultaneous synthesis of large numbers of oligodeoxyribonucleotides at low cost. Of course, this is the ideal way of preparing the overlapping fragments for a total gene synthesis, since by using a four column Omnifit synthesiser (*Figure 7*) up to a hundred or so oligodeoxyribonucleotides of about 15 bases in length can be assembled in a single day, a large proportion of the time being spent re-sorting the numbered discs at the end of each cycle of deoxyribonucleotide addition. Unfortunately at the time of writing it is our experience that this new method is not sufficiently reliable to be unequivocally recommended. We hope this situation will soon be remedied.

2.3 Protecting Groups

We favour the use of the 2-chlorophenyl (16) or 2,5-dichlorophenyl (17) protecting groups for the internucleotidic phosphate group since both are readily cleaved by oximate reagents (18) with a minimum of internucleotide bond cleavage.

The 9-phenylxanthen-9-yl (pixyl or Px) protecting group (19) is the favoured protecting group for the 5′-hydroxyl group, although the di-*p*-anisyl-phenylmethyl (dimethoxytrityl or DMTr; see *Figure 8*, Chapter 1) group is a perfectly acceptable alternative. The former group is preferred since it is more polar and confers crystallinity on many of the intermediates *en route* to the various monomers (thus avoiding some short column chromatography), and it is marginally more acid labile. The pixyl cation when liberated by acid is yellow-green in colour, whereas the dimethoxytrityl cation is a deep orange colour.

The standard benzoyl and isobutyryl groups are used for the protection of the N^4- and N^2-exocyclic amino groups of 2′-deoxycytidine and 2′-deoxyguanosine, respectively. Depurination of the conventionally used N^6-benzoyl-2′-deoxyadenosine is occasionally a problem (and particularly so when it is at the 3′ terminus of the chain) when protic acids are used for the removal of the 5′-protecting group. The recently developed N^6-phthaloyl (20,21) and N^6-di-n-butylaminomethylidene (22) groups are a considerable improvement. Almost no depurination is observed in either case but the latter is the protecting group of

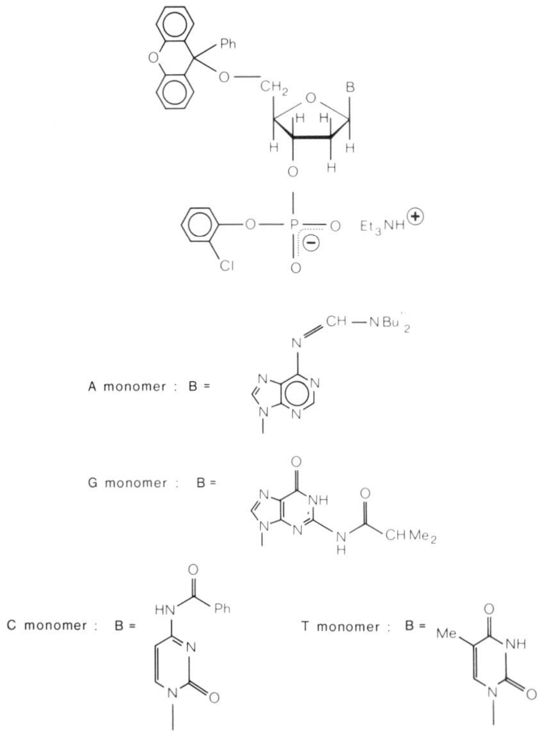

Figure 3. Structures of the four 5'-O-pixyl protected monomers carrying the preferred heterocyclic base-protecting groups.

choice. The structures of the four deoxyribonucleotide monomers, whose syntheses are dealt with in Sections 3.2.5 – 3.2.8, are illustrated in *Figure 3*. At the time of writing the N^6-di-n-butylaminomethylidene-protected 2'-deoxyadenosine derivative is not yet commercially available, neither are pixyl-protected monomers. The standard N^6-benzoyl-2'-deoxyadenosine derivative can still be used with good results in most syntheses of up to about 20 bases in length, especially where the adenine content is not unusually high.

2.4 Chain Assembly

Oligodeoxyribonucleotides are synthesised from the 3' end to the 5' end in a cyclical method involving just two chemical reactions per cycle (*Figure 4*). The first of these is the removal of the 5'-protecting group (DMTr or Px) by a suitable protic acid, namely 10% (w/v) trichloroacetic acid in 1,2-dichloroethane when the rather basic polydimethylacrylamide-Kieselguhr support is used or 3% (v/v) dichloroacetic acid in the above solvent when controlled pore glass is used. After appropriate solvent washes to remove any residual acid, the appropriately protected phosphodiester component (the monomer), as its triethylammonium salt, is condensed with the free 5'-hydroxyl group of the deoxyribonucleoside bound to the support in the presence of the coupling agent 1-mesitylenesulphonyl-3-

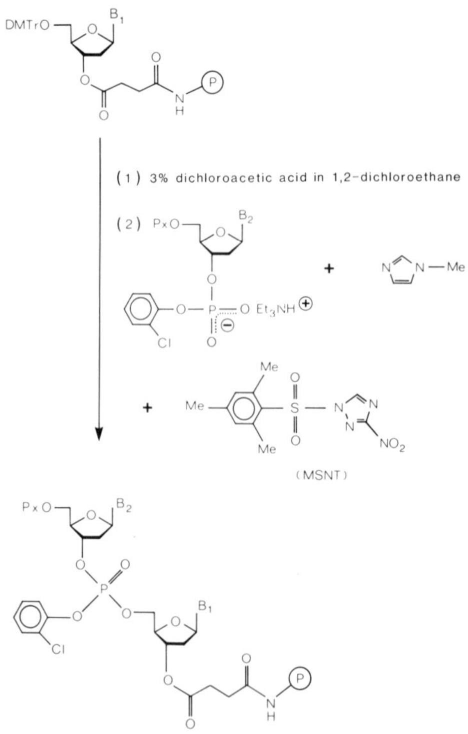

Phosphotriester Method

Figure 4. Reactions involved in a single cycle of chain assembly, for example cycle 1 to form a dimer as shown.

nitro-1,2,4-triazole (MSNT, see *Figure 4*) and the catalyst 1-methylimidazole in dry pyridine. In our experience a subsequent 'capping' step (treatment with a reagent such as acetic anhydride) has not been found necessary. We find no evidence for the presence of a significant amount of unreacted 5'-hydroxyl groups in normal circumstances after coupling reactions by the phosphotriester route. In our opinion truncated sequences must predominantly arise by a mechanism other than non-reaction (e.g., capping off by impurities during coupling, sulphonation, or internucleotide cleavage reactions).

In principle any of the types of assembly apparatus (manual, semi-manual, semi-mechanised or mechanised, see Chapter 1) could be used for this chemistry. We favour in general a semi-manual system which offers a useful compromise between manual and mechanised systems. Manual labour is reduced to a minimum, thus taking advantage of the original aim of solid-phase synthesis, yet the apparatus is relatively cheap and is flexible enough for multiple syntheses. The polymer support is contained in a small glass column connected to a manually operated solvent delivery system operated by a slight pressure of argon or nitrogen. We have utilised an inexpensive kit (Omnifit, Ltd., Cambridge, UK) which can be used to run several columns simultaneously. However, any reliable flow system could be adapted for this purpose and there are now other purpose built systems available. The cycle time for the small-scale procedure using controlled pore glass is about 24 min and involves relatively little manual work during chain assembly steps.

2.5 **Deprotection and Purification**

After the appropriate number of reaction cycles the support is treated with the following.

(i) 1,1,3,3-Tetramethylguanidinium *syn*-2-nitrobenzaldoximate (18) or pyridine-2-carbaldoximate in aqueous dioxan to remove the 2-chlorophenyl protecting groups and cleave the succinate linkage, thus liberating the partially deprotected oligodeoxyribonucleotide into solution. In addition, guanine O^6-modification by nitrotriazole, a known and allowed for side reaction (23) is reversed by prolonged oximate treatment.

(ii) The evaporated supernatant and washings from above are treated with concentrated aqueous ammonia (~35% ammonia by weight; specific gravity 0.88) in a sealed flask for several hours at 60°C to remove the acyl protecting groups. A 3–6 h treatment is normally sufficient when 2'-deoxyadenosine bears N^6-benzoyl protection but about 15 h treatment is needed to achieve removal of the di-n-butylaminomethylidene protecting group.

(iii) After evaporation of the solution from (ii) the residue is treated with acetic acid/water (8:2 v/v) for 20–30 min to remove the 5'-terminal protecting group.

Normally the crude material is then purified by ion-exchange h.p.l.c., followed by desalting either on a column of Bio-Gel P-2 eluted with ethanol/water (2:8 v/v) or by dialysis, and then further purified, if necessary, by reversed phase h.p.l.c. Purification by polyacrylamide gel electrophoresis can be used in place of ion-exchange h.p.l.c. Alternatively the crude oligodeoxyribonucleotide after the ammonia step can be purified by reversed phase h.p.l.c. prior to removal of the 5'-protecting group. The dimethoxytrityl group has a marked retarding effect on the elution of the target sequence bearing it. However, this effect is much less marked when the much more polar pixyl group is at the 5' terminus.

3. EXPERIMENTAL PROCEDURES

3.1 **Preparation of Supports**

3.1.1 *Polydimethylacrylamide-Kieselguhr*

To functionalise this support follow the procedure given in *Table 1*. Carry out the functionalisation in a glass reaction vessel fitted with a ground glass joint and stopper at the top and a sintered glass frit and a tap at the bottom (*Figure 5*). Add solvents and reagents batchwise through the ground glass joint and remove them by slight nitrogen pressure to the top of the vessel. Agitate the resin *gently* for the first 0.5–1 min of each wash only; avoid vigorous agitation as this will generate fines which will then block the sintered glass frit. Note that the resin in the free amino form sticks to the walls of the vessel. Use 5 ml of wash solution per gram of dry resin.

Phosphotriester Method

Table 1. Functionalisation of Polydimethylacrylamide-Kieselguhr.

Treatment	Number of washes	Time per wash (min)	Ninhydrin test[a]
Anhydrous ethylene diamine	1	16 h	
DMF wash	10–15	2	Resin + ve
			Last wash – ve
10% diisopropylethylamine in DMF wash	3	5	
DMF wash	5	2	
5 Equivalents of Fmoc-glycine anhydride[b] in DMF	1	90	
DMF wash	10	2	Resin – ve
Piperidine/DMF (2:8 v/v) wash	1	10	
DMF wash	5	2	Resin + ve
5 Equivalents of Fmoc-glycine anhydride[b] in DMF	1	90	
DMF wash	10	2	Resin – ve
You can either wash the resin with dichloromethane (5 x), dioxan (5 x) and diethyl ether (5 x) then dry it *in vacuo* and store it, or continue with the final step of the functionalisation, the attachment of the deoxyribonucleoside-3'-O-succinate			
Piperidine/DMF (2:8 v/v) wash	1	10	
DMF wash	5	2	Resin + ve
5 Equivalents of symmetrical anhydride of deoxyribonucleoside-3'-O-succinate[c] in DMF	1	90	
DMF wash	5	2	Resin – ve
Dichloromethane	5	2	
Dioxan	5	2	
Diethyl ether	5	2	
Dry the resin *in vacuo* and store it at $-20°C$[d]			

[a]Remove a few beads of resin from the reaction vessel and wash them on a sinter with DMF, dichloromethane, dioxan, and diethyl ether. Place a few of the dry beads in a small glass tube and assay by the ninhydrin test (see Appendix I). The test can also be carried out on one drop of eluent.

[b]Preparation of Fmoc-glycine symmetrical anhydride. Dissolve 9-fluorenylmethoxycarbonylglycine (10 equivalents over the sarcosine loading of the resin; mol. wt. 297.32) in a small volume of dichloromethane (~7 ml/g of glycine derivative) and add the minimum volume of DMF required to obtain solution. Add a solution of DCCI (5 equivalents over the sarcosine loading of the resin; mol. wt. 206.33. **Caution: use gloves and eye protection when handling DCCI**) in dichloromethane (~8 ml/g) and stir in a stoppered flask. A fine white precipitate of DCU will form within a few minutes. (N.B. It is our experience that some commercial batches of DCCI arrive already partially hydrolysed to DCU. Therefore, scrape off the surface layer and take only the translucent material underneath. It should dissolve easily in dichloromethane. Do not attempt to weigh out less than 100 mg. If less than this is required weigh out a convenient multiple of this amount, dissolve it in dichloromethane and add the requisite aliquot of the solution instead of solid DCCI. If only a little precipitate or no precipitate of DCU forms in the reaction it is probably due to insufficient unhydrolysed DCCI being present. DCCI is best purified by distillation at reduced pressure (see Section 3.3.5). After 15 min evaporate the mixture to dryness *in vacuo* (do not heat) and dissolve the residue in DMF (use 5 ml/g of resin to be treated), filter through a small sintered glass funnel and transfer the solution immediately onto the resin.

[c]Preparation of the symmetrical anhydride of deoxyribonucleoside-3'-O-succinates. Dissolve the dry protected deoxyribonucleoside-3'-O-succinate (10 equivalents over the sarcosine loading of the resin) in dry dichloromethane (~5 ml/g) and add a solution of DCCI (5 equivalents over the sarcosine

loading of the resin) in a small volume of dichloromethane. Stopper the flask and stir magnetically for 15 min. Work up as described for Fmoc-glycine symmetrical anhydride in Note b above. Molecular weights are: DMTrdbzA-3'-O-succinic acid, 757.8; DMTrdibG-3'-O-succinic acid, 739.8; DMTrdbzC-3'-O-succinic acid, 733.8; DMTrdT-3'-O-succinic acid, 644.7.

[d]Estimate dimethoxytrityl spectrophotometrically. Weigh accurately 2−3 mg of dry resin and place it in a 25 ml volumetric flask. Make up to the mark with 60% aqueous perchloric acid/ethanol (3:2 v/v) and shake for 5−10 min. Measure the absorbance of the orange solution at 495 nm in a spectrophotometer. Multiply the reading by 25 to give the total absorbance units. Assuming 1 µmol of dimethoxytrityl cation corresponds to 71.7 absorbance units at 495 nm, calculate the number of µmol of dimethoxytrityl group on the sample of resin weighed out and hence determine the number of µmol per gram, which is a measure of the deoxyribonucleoside loading.

3.1.2 *Long Chain Alkylamine Controlled Pore Glass (LCAA/CPG)*

(i) Dry long chain alkylamine controlled pore glass (1 g) *in vacuo* overnight over P_2O_5 and transfer it into a 10 ml capacity glass reaction vessel (*Figure 5*).

(ii) Wash the support with a 10% (v/v) solution of diisopropylethylamine in DMF (3 x 5 ml) for 5 min per wash (this ensures it is all in the free amino form) and then with DMF (5 x 10 ml).

(iii) During the DMF washes prepare 0.2 mmol of the symmetrical anhydride of one of the 2'-deoxynucleoside-3'-O-succinates as described in notes b and c of *Table 1*. However, do not filter off the dicyclohexylurea (DCU) but add the entire DMF mixture (use ~3 ml) onto the support and leave for 5−6 h.

(iv) Drain the support, wash it with DMF (5 x 10 ml) then 4 x 10 ml pyridine (note that the beads will give a positive ninhydrin test as well as a positive dimethoxytrityl test).

(v) To cap off the unreacted amino groups add a mixture of acetic anhydride/pyridine (5 ml, 1:9 v/v) to the support and leave for 1 h.

(vi) Wash the support with pyridine (5 x 10 ml), followed by diethyl ether (5 x 10 ml) and then dry it carefully *in vacuo*.

(vii) Carry out a dimethoxytrityl analysis on a few milligrams of the support to determine the loading (varies between 10 and 30 µmol/g) according to batch and store the dried material at −20°C.

3.2 Preparation of Deoxyribonucleotide Monomers, Deoxyribonucleoside-3'-O-succinates and Coupling Agent

3.2.1 *Preparation of Di-n-butylformamide Dimethyl Acetal*

(i) Heat di-n-butylamine (21 ml, 125 mmol) and N,N-dimethylformamide dimethyl acetal (18.26 ml, 137.5 mmol) at 100°C (oil-bath) for 3 days in a round-bottomed flask equipped with a reflux condenser and drying tube containing calcium chloride.

(ii) When cool, fractionally distil the yellow mixture under dry nitrogen at about 8−10 mm Hg pressure (see Appendix I) to obtain di-n-butylformamide dimethyl acetal as a colourless liquid boiling at about 84−86°C at 8 mm Hg pressure.

(iii) Store the liquid in a dark bottle under nitrogen.

Phosphotriester Method

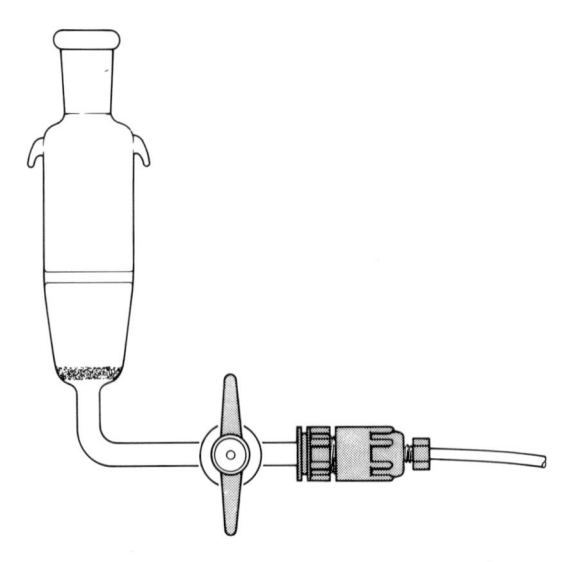

Figure 5. Vessel used for resin functionalisation; a larger capacity version (250 ml) is used for the preparation of 2-chlorophenyl phosphorodi(triazolide).

3.2.2 *Preparation of N^6-di-n-butylaminomethylidene-2'-deoxyadenosine*

(i) Dissolve 2'-deoxyadenosine monohydrate (6.73 g, 25 mmol) in dry DMF (250 ml) and evaporate the solution to dryness *in vacuo* (use a rotary evaporator and an oil pump).

(ii) Dissolve the residue in dry DMF (160 ml) and add di-n-butylformamide dimethyl acetal (7.5 ml, 32.1 mmol).

(iii) Fill the flask with argon, stopper, and stir the pale yellow solution for 24 h at room temperature in the dark.

(iv) Evaporate the solution *in vacuo* to leave a pale yellow oil.

(v) Dissolve this oil in dichloromethane (300 ml) and shake vigorously with 5% aqueous sodium bicarbonate solution (300 ml) in a separating funnel (with the funnel stoppered and in an inverted position release any pressure every now and then by opening and closing the tap) for about 1 min.

(vi) Stand the flask upright so that the layers separate completely.

(vii) Remove the lower (organic) layer and pour the aqueous layer to waste.

(viii) Repeat the washing once more, remove the organic layer and dry it by adding some anhydrous sodium sulphate.

(ix) Allow to stand for 10 min, filter the mixture and evaporate the filtrate *in vacuo* to leave a yellow syrup.

(x) Silica gel t.l.c. of the product run in ethanol/chloroform (1:9 v/v) containing 0.5% pyridine should show an intense u.v.-positive spot of R_f about 0.25.

(xi) Spray the plate in a fume cupboard with 60% aqueous perchloric acid/ethanol (3:2 v/v) and heat it for a few minutes at 90°C in an oven and the spot will char.

The crude product can now be used directly in the next reaction.

3.2.3 Preparation of 5'-O-pixyl-N^6-di-n-butylaminomethylidene-2'-deoxyadenosine

(i) Dry the syrupy N^6-di-butylaminomethylidene-2'-deoxyadenosine (~25 m-mol, from Section 3.2.2) by evaporation of dry pyridine (2 x 50 ml) *in vacuo*.

(ii) Dissolve the residual oil in dry pyridine (60 ml) and add 9-chloro-9-phenyl-xanthene (8.78 g, 30 mmol; pixyl chloride) with magnetic stirring and exclusion of moisture (stopper the flask), and keep the flask in the dark. The solution will probably turn deep green in colour.

(iii) Monitor the course of the reaction by silica gel t.l.c. in 10% ethanol/90% chloroform containing 0.5% pyridine.

(iv) Examine the dry plate under u.v. light (**eye protection required**) and then spray the plate with 60% aqueous perchloric acid/ethanol (3:2 v/v) and heat for a few minutes. Compounds bearing pixyl groups will give dark yellow/green spots. You will probably observe several pixyl-positive spots (the desired product has an R_f of about 0.57).

(v) Terminate the reaction by addition of methanol (19 ml) when all the starting material (R_f 0.22 – 0.25) has been consumed; this takes about 60 min.

(vi) Partition the solution between chloroform (150 ml) and saturated aqueous sodium bicarbonate solution (200 ml).

(vii) Separate the organic layer (lower) and wash the aqueous phase with chloroform (2 x 80 ml).

(viii) Dry the combined organic layers over anhydrous sodium sulphate, filter the mixture, and evaporate the filtrate *in vacuo* (do not heat above 35°C).

(ix) Dry the residual orange syrup by addition and evaporation *in vacuo* of dry pyridine (2 x 50 ml).

(x) Next, remove any residual pyridine by co-evaporation with dry toluene (50 ml) to leave the crude product as a yellow foam.

(xi) Dissolve the resultant foam in dichloromethane (100 ml) containing 0.5% triethylamine (to prevent loss of the pixyl group) and chromatograph the solution on a short Kieselgel 60H column (480 g of silica in a bed 13 cm wide by about 10.2 cm deep) using about 2.5 p.s.i. of nitrogen pressure (see Appendix I).

(xii) Elute the column with dichloromethane/0.5% triethylamine (~30 ml) followed by 2 litres of 3% ethanol in dichloromethane/0.5% triethylamine and then 2 litres of 4% ethanol in dichloromethane/0.5% triethylamine.

(xiii) Collect fractions of about 25 ml and monitor them by $A_{260\,nm}$ and by silica gel t.l.c. in 10% ethanol/90% chloroform containing 0.5% pyridine. The desired product has an R_f of about 0.57.

(xiv) Pool those fractions containing pure product as determined by t.l.c. and evaporate the solution *in vacuo* to leave a glass.

(xv) Dry the usually pale yellow glass by evaporation of pyridine *in vacuo* and finally remove residual pyridine by co-evaporation with dry toluene (50 ml).

Phosphotriester Method

(xvi) Dissolve the residual glass (or foam) in dichloromethane (~16 ml) and precipitate the product by dropwise addition of the solution to vigorously stirred dry pentane (700 ml; previously dried by passage through a column of basic alumina).

(xvii) Collect the white precipitate by centrifugation at about 1200 r.p.m. at 4°C, decant off the supernatant and wash the precipitate with pentane (2 x 700 ml) containing about 0.5% of triethylamine, centrifuging and decanting after each wash. (**Warning: use capped tubes because pentane has a low flash point**).

(xviii) Finally, air dry the product and then dry it carefully in a vacuum desiccator (apply the vacuum gradually, otherwise the pentane wet material will sputter) over separate containers of potassium hydroxide pellets (KOH) and phosphorus pentoxide (P_2O_5). You should obtain about 9.6 g of product; about 60% yield.

(xix) Store the dry product in a tightly closed container at $-20°C$.

3.2.4 *Preparation of 2-chlorophenyl Phosphorodi(triazolide) (Active Phosphorylating Agent)*

(i) Dissolve dry 1,2,4-triazole (2.76 g, 40 mmol) and dry triethylamine (4.87 ml, 35 mmol) in dry freshly distilled tetrahydrofuran (THF, 100 ml) in a 250 ml capacity glass vessel (*Figure 5*).

(ii) Add 2-chlorophenyl phosphorodichloridate (2.42 ml, 15 mmol) and shake the mixture under anhydrous conditions (stoppered) for 30 min. A copious white precipitate of triethylammonium chloride forms immediately.

(iii) Transfer the mixture to a flask containing pre-dried 5'-protected base-protected 2'-deoxynucleoside for the phosphorylation reaction.

3.2.5 *Preparation of Triethylammonium[5'-O-pixyl-N^6-di-n-butylaminomethylidene-2'-deoxyadenosine-3'-O-(2-chlorophenyl Phosphate)]*

(i) Dry 5'-O-pixyl-N^6-di-n-butylaminomethylene-2'-deoxyadenosine (6.47 g, 10 mmol) by evaporation of dry pyridine (30 ml) and then transfer a freshly prepared solution of 2-chlorophenyl phosphorodi(triazolide) (15 mmol, prepared as in Section 3.2.4) in dry THF into the flask of protected deoxyribonucleoside.

(ii) Carry out the transfer, through Teflon tubing connected to the Altex fitting of the vessel, by applying nitrogen pressure to the top of the glass reaction vessel, and then wash the triethylammonium chloride retained on the glass sinter with a little dry THF (2 x 10 ml).

(iii) Leave the solution under anhydrous conditions (stoppered flask) at room temperature and monitor the progress of the reaction by silica gel t.l.c. in ethanol/chloroform (1:9 v/v) containing 0.5% pyridine.

(iv) When no starting material (R_f ~0.57) remains you will observe a pixyl-positive spot on the baseline – this generally takes about 60–80 min – terminate the reaction by addition of aqueous triethylammonium bicarbonate (200 ml, 1 M, pH ~8).

(v)   After 5 min extract the solution with chloroform or dichloromethane (300 ml).
(vi) Separate the pale yellow organic layer and wash it with 0.5 M triethylammonium bicarbonate solution (2 x 300 ml) followed by water (300 ml).
(vii) Dry the organic layer over sodium sulphate, filter the mixture after 5 min and evaporate the filtrate *in vacuo* in the presence of pyridine (~ 10 ml).
(viii) Dry the residual oil by evaporation of dry pyridine (50 ml), and finally remove residual pyridine by co-evaporation with dry toluene (50 ml), leaving the crude product as a foam.
(ix) Dissolve the residual foam in dichloromethane (30 ml) containing 1% triethylamine and chromatograph the solution on a short Kieselgel 60H column (130 g of silica in a bed 7.5 cm wide by about 8.5 cm deep) using about 2.5 p.s.i. of nitrogen pressure.
(x)   Elute the column with 250 ml of 4% ethanol in dichloromethane containing 1% triethylamine followed by ethanol/dichloromethane/triethylamine (1 litre, 7:92:1 by volume).
(xi) Collect fractions of about 20 – 25 ml and monitor them by $A_{260 \text{ nm}}$ and by silica gel t.l.c. in ethanol/chloroform (1:9 v/v) containing 0.5% pyridine.
(xii) Pool those fractions containing pure product (R_f about 0.06 – 0.10) and evaporate the solution *in vacuo* to leave a white foam.
(xiii) Dissolve this foam in dichloromethane (20 ml) containing 1% triethylamine and precipitate the product by dropwise addition of the solution to vigorously stirred dry pentane (700 ml) containing 1% triethylamine.
(xiv) Collect the white precipitate by centrifugation at 4°C wash it with pentane (2 x 700 ml), air dry it, and then dry it carefully *in vacuo* over P_2O_5 and KOH (separate containers).
(xv) Store the dry product (a monomer – see *Figure 3* for structure) (it is slightly hygroscopic) in a sealed container at −20°C. You can expect to obtain about 8.2 g (73% yield).

3.2.6 *Preparation of Triethylammonium[5'-O-pixyl-N²-isobutyryl-2'-deoxyguanosine-3'-O-(2-chlorophenyl Phosphate)]*

(i)   Prepare (or buy) 5'-O-pixyl-N²-isobutyryl-2'-deoxyguanosine according to the procedure given in Chapter 2 (but replacing DMTr chloride by the requisite amount of pixyl chloride) and crystallise the product from acid-free chloroform (purified by passage through a column of basic alumina).
(ii) Dry the compound (5.94 g, 10 mmol) by evaporation of dry pyridine (50 ml) then add dry pyridine (10 ml) followed by 15 mmol of 2-chlorophenyl phosphorodi(triazolide) and then follow the procedure given in Section 3.2.5. The desired product (G monomer – see *Figure 3*) has an R_f of 0.03 on silica gel t.l.c.
(iii) The workup and isolation of product is as described in Section 3.2.5. You can expect to obtain about 6.6 g (75%) of product as a practically white powder, which should be stored at −20°C.

3.2.7 Preparation of Triethylammonium[5'-O-pixyl-N⁴-benzoyl-2'-deoxycytidine-3'-O-(2-chlorophenyl Phosphate)]

(i) Prepare (or buy) 5'-O-pixyl-N⁴-benzoyl-2'-deoxycytidine (R_f ~0.35 on silica gel t.l.c. in the usual ethanol/chloroform system) according to the procedure given in Chapter 2, but substituting pixyl chloride for DMTr chloride, and crystallise the material from dry benzene (distil from P_2O_5 — **Warning: very toxic, use fume hood**). You require 5.88 g for a 10 mmol scale reaction.

(ii) Follow the procedure as detailed in Section 3.2.5.

(iii) The desired product (C monomer — see *Figure 3*) has an R_f of about 0.03 on silica gel t.l.c. in ethanol/chloroform (1:9 v/v). You can expect to obtain about 6.5 g (75% yield) of product as a white powder.

(iv) Store the dry product at −20°C.

3.2.8 Preparation of Triethylammonium[5'-O-pixyl thymidine-3'-O-(2-chlorophenyl Phosphate)]

(i) Prepare (or buy) 5'-O-pixyl thymidine according to the procedure of Reese (19) and recrystallise it from benzene (**Warning: toxic**). The product has an R_f of 0.43 on silica gel t.l.c. run in ethanol/chloroform (1:9 v/v) containing 0.5% pyridine. Use 4.99 g of material for a 10 mmol scale reaction.

(ii) Follow the phosphorylation and purification procedure given in Section 3.2.5. The desired product (T monomer — see *Figure 3*) has an R_f of about 0.04. You can expect to obtain ~6.3 g (80% yield) of product as a white powder.

(iii) Store the dried material at −20°C.

3.2.9 General Procedure for the Preparation of 5'-O-dimethoxytrityl-2'-deoxyribonucleoside-3'-O-succinates

(i) Dissolve the dry protected 2'-deoxyribonucleoside (4 mmol; 2.18 g of DMTrdT, 2.53 g of DMTrdbzC, 2.63 g of DMTrdbzA, or 2.56 g of DMTrdibG) in dry DMF (25 ml) and add 4-dimethylaminopyridine (1.18 g, 0.89 mmol) followed by succinic anhydride (0.99 g, 9.82 mmol).

(ii) Leave the mixture overnight at room temperature in the dark. T.l.c. should show complete reaction (see *Table 2* for t.l.c. data on protected nucleosides and their 3'-O-succinates).

(iii) Evaporate the solution *in vacuo* to an oil.

(iv) Dissolve the oil in pyridine/water (100 ml, 4:6 v/v) and add about 200 ml of Dowex 50-X8 resin (pyridinium form).

(v) Leave for 10 min then pour the mixture into a column of pyridinium Dowex resin (~200 ml) and elute slowly with pyridine/water (~1 litre, 4:6 v/v).

(vi) Evaporate the solution to dryness in the presence of pyridine, dry by evaporation of dry pyridine (2 x 50 ml) and then remove residual pyridine by co-evaporation with dry toluene (50 ml).

(vii) Dissolve the residual foam in dichloromethane (~20 ml) containing 0.5% pyridine and apply the solution, after filtration, to a short Kieselgel 60H col-

Table 2. R_fs of Protected 2'-Deoxyribonucleosides and their 3'-O-Succinates in Two Different Solvent Systems used with Silica Gel Plates.

Compound	R_f on silica gel t.l.c. in ethanol/chloroform (1:9) containing 0.5% pyridine	R_f on silica gel t.l.c. in ethyl acetate/acetone/water (5:10:1 v/v) containing 0.5% pyridine
DMTrdbzA	0.47	0.70
DMTrdibG	0.29	0.64
DMTrdbzC	0.52	0.66
DMTrdT	0.50	0.74
DMTrdbzA-3'-O-succ.	0.24*	0.52
DMTrdibG-3'-O-succ.	0.11*	0.46
DMTrdbzC-3'-O-succ.	0.28*	0.53
DMTrdT-3'-O-succ.	0.24*	0.59

*Spots show pronounced tailing in this solvent system.

umn (100 g of silica in a bed about 7.5 cm wide by 7 cm deep). Elute the column with about 0.5 litres of dichloromethane containing 0.5% pyridine followed by about 1 litre of 4 – 5% ethanol in dichloromethane containing 0.5% pyridine.

(viii) Monitor the fractions by $A_{260\ nm}$ and by silica gel t.l.c. in the two solvent systems given in *Table 2*.

(ix) Pool only those fractions containing pure product as determined by t.l.c. (reject any fractions containing higher or lower R_f spots) and evaporate them to dryness *in vacuo*.

(x) Dry the residual glass by evaporation of dry pyridine (2 x 25 ml) and then dissolve the residue in dichloromethane (25 ml) and precipitate the product by dropwise addition of this solution to vigorously stirred pentane/diethyl ether (500 ml; 1:1 v/v).

(xi) Collect the product by centrifugation and wash with pentane/ether (2 x 50 ml; 1:1 v/v).

(xii) Dry the product carefully *in vacuo* over P_2O_5 and KOH and store at −20°C. Yield about 80%.

3.2.10 *Preparation of 3-nitro-1,2,4-triazole*

(i) Place 3-amino-1,2,4-triazole (75 g, 0.9 mmol), sodium nitrite (300 g, 4.35 mol) and water (450 ml) in a 2 litre beaker.

(ii) Stir the mixture mechanically, cool the beaker in an ice-salt bath, and add concentrated nitric acid (255 ml) dropwise during 5 h while maintaining the temperature between 0° and 10°C. (Exercise care during the addition as there will be considerable frothing and evolution of heat and nitrous fumes.)

(iii) Allow the thick yellow mixture to warm to room temperature overnight, then cool it to 0°C (in ice) and collect the pale yellow solid by filtration at the pump.

(iv) Wash the solid with a little ice-cold water (3 x 150 ml), suck it as dry as possible, then dry it thoroughly *in vacuo* over P_2O_5.
(v) Recrystallise the crude product from about 2 litres of ethyl acetate (beware of using too much solvent since there will be a considerable amount of insoluble material which will need to be removed by filtration) to obtain large pale yellow plates. You can expect to obtain about 54 g of product (53% yield).

3.2.11 *Preparation of 1-mesitylenesulphonyl-3-nitro-1,2,4-triazole (MSNT)*

(i) Suspend mesitylenesulphonyl chloride (43.5 g, 0.2 mol, recrystallised from dry pentane) and finely powdered 3-nitro-1,2,4-triazole (22.8 g, 0.2 mol) in dry freshly distilled dioxan (625 ml) in a 1 litre round-bottomed flask cooled on ice.
(ii) Add dropwise during 10 min a solution of dry triethylamine (27.8 ml, 0.2 mol) in dioxan (40 ml) with stirring and exclusion of moisture (use a pressure equalising dropping funnel equipped with a drying tube).
(iii) Allow the mixture to warm to room temperature and continue stirring it for 1 h.
(iv) Filter off the precipitate of triethylammonium chloride at the pump, wash it with dry dioxan (3 x 30 ml), and evaporate the combined filtrates to dryness *in vacuo*.
(v) Recrystallise the pale yellow solid from about 120 ml of hot dry dioxan to obtain pale yellow cubic crystals (you can expect ~50 g, 90%), m.p. 135.5 – 137°C. The structure of MSNT is shown in *Figure 5*.

3.3 Purification of Solvents and Reagents

3.3.1 *2-Chlorophenyl Phosphorodichloridate*
(i) Vacuum distil before use using an oil pump and a dry nitrogen bleed (Appendix I).
(ii) Discard the first 10% of the distillate and collect the main fraction boiling at about 120°C at 3 mm Hg pressure.
(iii) Store the colourless liquid in a sealed flask preferably under nitrogen. **Warning: the liquid is corrosive and very moisture-sensitive.**

3.3.2 *1,2-Diaminoethane (Ethylene Diamine)*
(i) Reflux with sodium metal (**Danger: fire hazard**) (10 – 15 g per litre) and then distil under nitrogen at atmospheric pressure.
(ii) Discard the first few percent of the distillate and collect the main fraction boiling at 117°C at 760 mm Hg pressure.
(iii) Store the liquid in a dark bottle.

3.3.3 *1,2-Dichloroethane*
(i) Pass through a column of basic alumina (~100 g per litre) and then distil from phosphorus pentoxide (~5 g per litre) at atmospheric pressure.

(ii) Discard the first few percent of the distillate and collect the main fraction boiling at 83°C at 760 mm Hg pressure.
(iii) Store in a stoppered flask, without additives. **Warning: harmful vapour.**

3.3.4 *Dichloromethane*
(i) Distil from phosphorus pentoxide (**Danger: corrosive**) (5 – 10 g per litre) at atmospheric pressure after a short reflux.
(ii) Discard the first few percent of the distillate and collect the main fraction boiling at 40°C at 760 mm Hg pressure.
(iii) Store as for dichloroethane. **Warning: harmful vapour.**

3.3.5 *N,N'-Dicyclohexylcarbodiimide (DCCI)*
(i) Vacuum distil cautiously under dry nitrogen using an oil pump.
(ii) Discard the first few percent of the distillate and collect the main fraction boiling at about 120°C at 0.7 mm Hg pressure.
(iii) Store the waxy solid in a sealed dark bottle in the fridge. **Warning: DCCI is highly dangerous in contact with skin; contact with eyes can lead to blindness. Wear gloves and eye protection at all times. Do not breathe vapour.**

3.3.6 *N,N-Diisopropylethylamine*
(i) Distil initially from ninhydrin (1 – 2 g per litre) and then redistil from potassium hydroxide pellets under dry nitrogen at atmospheric pressure.
(ii) Discard the first 10% of the distillate and collect the main fraction which boils at 127°C at 760 mm Hg pressure.
(iii) Store the colourless liquid in dark bottles, without additives.

3.3.7 *N,N-Dimethylformamide (DMF)*
(i) Fractionally distil analytical grade DMF under reduced pressure (water pump) using a dry nitrogen bleed and a fractionating column filled with glass helices (see Appendix I).
(ii) Discard the first 10% of the distillate and collect the main fraction boiling at about 50°C at 15 mm Hg pressure.
(iii) Do not store over molecular sieves and distil just prior to use because the pure material starts to decompose within a few days. **Warning: DMF causes harm if it is inhaled or absorbed through the skin.**

3.3.8 *4-Dimethylaminopyridine*
Recrystallise from diethyl ether containing a little decolourising charcoal, to obtain colourless plates, m.p. 110 – 112°C. N.B. The charcoal must be removed by filtering the hot solution through Celite before crystallisation occurs.

3.3.9 *1,4-Dioxan*
(i) Pass through a column of basic alumina (~200 g per litre) to remove peroxide impurities and most of the water present.

Phosphotriester Method

(ii) If totally anhydrous dioxan is required reflux with sodium metal (**Warning: fire hazard**) and benzophenone until the deep mauve-blue colour of the ketyl radical appears, indicating that the solvent is anhydrous.
(iii) Then distil at atmospheric pressure discarding the first few percent of the distillate and collect the main fraction boiling at 101°C at 760 mm Hg pressure. Since the dry solvent soon forms peroxides only distil it when required for immediate use.
(iv) Store in dark bottles. **Warning: harmful vapour.**

3.3.10 *1-Methylimidazole*

(i) Vacuum distil under dry nitrogen using an oil pump.
(ii) Discard the first few percent of the distillate and collect the main fraction boiling at about 65°C at 4 mm Hg pressure.
(iii) Store the colourless liquid in small vials under nitrogen or argon.

3.3.11 *Phenyl Isocyanate*

(i) Vacuum distil (water pump) in a fume hood using a dry nitrogen bleed and collect the main fraction boiling at 55°C at 13 mm Hg pressure.
(ii) Store the colourless liquid in a sealed flask at 4°C. **Warning: exercise great care when handling this compound as it is extremely toxic.**

3.3.12 *Piperidine*

(i) Distil from potassium hydroxide pellets (10–20 g per litre) under dry nitrogen at atmospheric pressure.
(ii) Discard the first few percent of the distillate and collect the main fraction boiling at 106°C at 760 mm Hg pressure.
(iii) Store in dark bottles. **Warning: harmful vapour.**

3.3.13 *Pyridine*

(i) Reflux with ninhydrin (~5 g per litre) for 2–3 h and then distil at atmospheric pressure. This removes any ammonia and primary or secondary amines present.
(ii) Finally reflux with barium oxide (~20 g per litre) or potassium hydroxide pellets for 2–3 h and fractionally distil at atmospheric pressure using a Vigreux column (see Appendix I) — barium oxide is the better drying agent.
(iii) Discard the first 5–10% of the distillate and collect the main fraction boiling at 116°C at 760 mm Hg pressure.
(iv) Store in a stoppered flask without additives.

Do not distil pyridine from calcium hydride since this results in the formation of secondary amines. Check for the presence of amines by the ninhydrin test (Appendix I). **Warning: harmful if inhaled or absorbed through skin.**

3.3.14 *Tetrahydrofuran (THF)*

(i) Pass through a column of basic alumina (~200 g per litre) to remove peroxide impurities and most of the water.

(ii) Then reflux with sodium metal (**Danger**) and benzophenone until the deep blue colour of the benzophenone ketyl radical appears, indicating that the solvent is anhydrous, and distil at atmospheric pressure.
(iii) Discard the first few percent of the distillate and collect the main fraction boiling at 65°C at 760 mm Hg pressure.
(iv) Store in the dark under nitrogen. **Warning: harmful vapour.** The pure, unstabilised solvent rapidly forms peroxides therefore only distil when necessary and use within 2 days.

3.3.15 *1,2,4-Triazole*

Recrystallise from dry dioxan in the presence of a little decolourising charcoal to obtain white needles, m.p. 120 – 121°C after drying *in vacuo*. (See note in Section 3.3.8.)

3.3.16 *Triethylamine*

(i) Distil from either potassium hydroxide pellets or sodium borohydride at atmospheric pressure under anhydrous conditions.
(ii) Discard the first few percent of the distillate and collect the main fraction boiling at 89°C at 760 mm Hg pressure.
(iii) Do not store over molecular sieves. **Warning: the vapour is irritating to skin, eyes and respiratory system.**

4. SYNTHESIS OF FULLY PROTECTED OLIGODEOXYRIBONUCLEOTIDES

4.1 Setting up a Manual Synthesis Apparatus

A schematic diagram of the apparatus is shown in *Figure 6*, and a photograph of a 4-column set up is shown in *Figure 7*. For the most part the setting up is com-

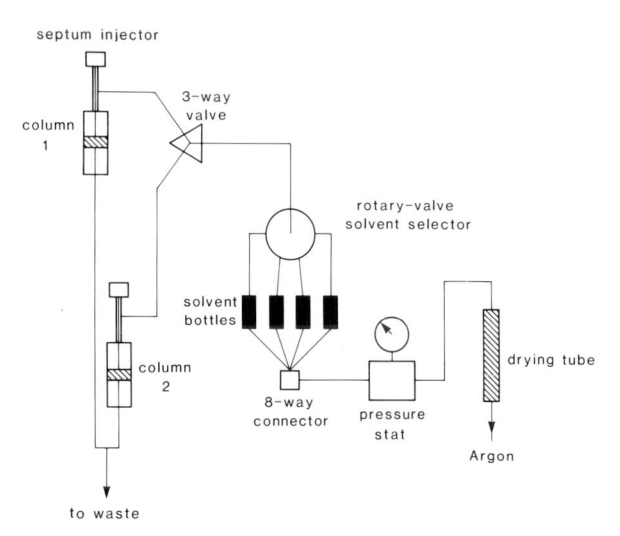

Figure 6. Schematic diagram of the semi-manual synthesis apparatus.

Phosphotriester Method

Figure 7. Photograph of a four-column Omnifit set up which allows simultaneous synthesis of four oligodeoxyribonucleotides.

mon sense. However, take note of the following points.
(i) Use thick Teflon tubing (1.5 mm i.d.) for all argon (or nitrogen) filled lines. Use 0.8 mm i.d. Teflon tubing for the solvent delivery lines from the tops of the solvent reservoirs to the rotary valve and also for the delivery line from the rotary solvent selector valve to the column inlet. Use about 0.5 m of 0.3 mm i.d. tubing from the colum outlet to the waste bottle in order to generate a slight back pressure. You will need to cut all tubing to the desired length and then slip on the threaded bolts plus the appropriate gripper fittings (these are Teflon washers fitted with a 316 grade stainless steel backing ring. Beware of other grades of stainless steel as corrosion may occur in the acid line liberating ferric ions which will cause degradation of protected oligodeoxyribonucleotides).
(ii) Note that all liquid flow fittings must be Teflon or equivalent, whereas the gas flow connectors can be made with Viton 'O' rings.

The general setting up procedure is as follows.
(i) Use the long column to make a silica gel drying tube and connect this between the argon (or nitrogen) supply and the inlet valve of the pressure regulator (this has an on/off switch and a screw-in needle valve).

(ii) Connect the outlet valve of the regulator to one position on an 8-way connector, then connect four positions of this connector to one terminal on each of the four bottle tops [three reservoirs are 1 litre, to be used for pyridine, 1,2-dichloroethane and DMF (not required for glass beads method), and one is either 100 or 200 ml for use with the deprotecting agent in 1,2-dichloroethane]. One outlet terminal on each bottle top will be used as a vent to enable the head space above the solvents to be flushed with argon or nitrogen when the bottles are filled.
(iii) Next, connect one terminal from each bottle to the solvent selector valve and fit a length of tubing to the underside of these terminals such that the tubing reaches the bottom of the reservoirs.
(iv) Connect the 6-way rotary solvent selector such that the solvent order is (either clockwise or anti-clockwise) stop; pyridine; 1,2-dichloroethane; dichloroacetic (glass beads method) or trichloroacetic acid (polyamide-Kieselguhr method) in dichloroethane; stop; DMF (not required for glass beads).
(v) Blank off the stop positions using the screw-in connectors.
(vi) Finally, connect the outlet from the solvent selector to the column inlet and connect 0.5 m of 0.3 mm i.d. Teflon tubing between the column outlet and the waste bottle (use a spare 1 litre bottle) *via* a 2-way polypropylene coupling fitted with an on/off key.

If you wish to use more than one column, connect the appropriate multi-way connector between the outlet position of the solvent selector and the various column inlets. Until you are familiar with the manipulations involved in an oligodeoxyribonucleotide assembly it is best to use only the single column set up. When you are reasonably experienced, four colour-coded columns can easily be operated simultaneously.

Before filling any of the solvent bottles check the systems for leaks by putting some dichloromethane into each bottle, pressurising each in turn, and checking the flow through the entire system with a column in position. Tighten any leaky liquid-liquid connectors and then dry out the bottles and tubing by blowing dry nitrogen through them. Fit the column with a Teflon faced silicone rubber septum in the screw cap on the column top fitting such that the Teflon face is on the *liquid side* and place a Teflon frit into the glass column tube such that it rests on top of the adjustable plunger bottom fitting.

4.2 Preparing the Materials for a Small-scale Synthesis on Controlled Pore Glass

(i) Normally, use 25 mg of the appropriate functionalised long chain alkylamine controlled pore glass (CPG) support (~ 0.75 μmol of bound 3'-terminal nucleoside when the loading is 30 μmol/g).
(ii) For each cycle of nucleotide addition you will need 20 mg of MSNT, 13.3 μmol of nucleotide monomer, 10 μl of 1-methylimidazole and 100 μl of anhydrous pyridine. Weigh out the requisite amounts of each of the four monomers required for the entire assembly into four labelled 5 ml pear-shaped flasks (B14 neck) or 3 or 5 ml microvials. You can use either the

Phosphotriester Method

Table 3. Molecular Weights of the Triethylammonium Salts of the Various Monomers Available.

Monomer	5'-O-DMTr compound	5'-O-pixyl compound
A	949.46[a]	938.50[b]
G	931.44	885.35
C	925.43	879.34
T	836.33	790.24

[a]N6-benzoyl protected.
[b]N6-di-n-butylaminomethylidene protection.

5'-O-pixyl protected monomers prepared as described in Section 3.2, which gives the best results, or the 5'-O-dimethoxytrityl protected monomers that can be home-made or obtained commercially. Refer to *Table 3* for the molecular weights of the various monomers as their triethylammonium salts [N.B. all carry 3'-O-(2-chlorophenyl phosphate) groups; dC carries N4-benzoyl and dG carries N2-isobutyryl protection).

(iii) Fit each flask with a silicone rubber 'Subaseal' septum (or cap the microvial with a silicone septum and screw cap) and insert a small syringe needle in each. If using flasks stand them in small beakers and place the flasks or microvials in a vacuum desiccator over P_2O_5 and KOH.

(iv) Weigh out 25 mg of the appropriate glass support in a small vial and place this in the desiccator as well. Evacuate the desiccator using an oil pump and leave overnight.

(v) Weigh out 20 mg ($\pm 1-2$ mg) of MSNT into each of *n* pre-dried 1 ml capacity microvials fitted with screw caps and Teflon faced silicone rubber septa where *n* is the number of cycles required to assemble the desired oligodeoxyribonucleotide (ensure that the Teflon lining faces the inside of the vial). Do not expose the MSNT to the open atmosphere for longer than necessary during the weighing.

(vi) Fill the appropriate solvent and reagent reservoirs with pyridine, 1,2-dichloroethane and a 3% by volume solution of dichloroacetic acid in dichloroethane respectively (the DMF reservoir is not needed for syntheses on glass supports), and flush the head space of each bottle in turn with argon (or nitrogen) by pressurising the bottle and opening the vent. Leave for several minutes at about 6 p.s.i. pressure and then close the vent.

4.3 Assembly of the Fully Protected Support-bound Oligodeoxyribonucleotide on Small Scale on CPG

(i) Release the vacuum in the desiccator containing monomer and support by introduction of dry nitrogen, and immediately remove the syringe needles from the septa.

(ii) Dissolve each monomer in the correct volume of anhydrous pyridine (volume required = y x 100 μl where y is the number of additions of that particular monomer) using a 1 ml gas-tight Hamilton syringe, and then

Table 4. Cycle Operations for Small-scale Synthesis on Controlled Pore Glass.

Step	Operation	Time (min)
1	Pyridine wash	2
2	1,2-dichloroethane (DCE) wash	1.5
3	Deprotection with 3% DCA in DCE	0.5 – 1.25[a]
4	DCE wash	1.5
5	Pyridine wash	2.5
6	Coupling (stop flow)	15

[a]The time depends on the base being deprotected as well as the 5'-O-protecting group. This step should be curtailed as soon as all the yellow-green (pixyl cation) or orange (dimethoxytrityl cation) colour has been washed from the support. In general, removal of pixyl is faster than removal of dimethoxytrityl, and 5'-dC and T are deprotected more slowly than 5'-dA or G.

cover the septum with Parafilm or Nescofilm to ensure that the needle holes are sealed, thus reducing the chance of water getting into the flasks.

(iii) Carefully transfer the CPG support into the Omnifit column so that it rests on the Teflon frit (use a small funnel).

(iv) Disconnect the solvent inlet tube from the column top piece and run each solvent through it in turn for about 10 sec to remove any air and purge all the solvent lines (remember to pressurise the solvent reservoirs to about 6 p.s.i. and to open all the appropriate taps) in the order pyridine, dichloroethane, dichloroacetic acid in dichloroethane, dichloroethane, and finally pyridine.

(v) Stop the flow on the solvent selector valve and reconnect the inlet tube to the column top.

(vi) Add dry pyridine to the column to within about 1 cm of the top with the column outlet closed and stir the support gently with the tip of a Pasteur pipette to dislodge any entrapped air bubbles.

(vii) Push in the column top piece with the screw cap removed and wipe away the pyridine that exudes from it.

(viii) Fit the top pinch clamp (the bottom one securing the adjustable plunger should already be in place and the plunger locked) and then screw on the column top containing the Teflon-lined septum. Since the support has already been dried a drying step with 10% phenyl isocyanate in pyridine is rendered unnecessary.

(ix) Start the first cycle using the times given in *Table 4* and adjust the flow-rate to about 1 ml per minute in pyridine by altering the pressure regulator as necessary.

(x) During the first pyridine wash, adjust the bottom plunger so as to leave about 1 mm height of liquid space above the surface of the support.

(xi) During the deprotection, step 3, rinse a 250 µl gas-tight Hamilton syringe with dry pyridine and as soon as you commence step 4, withdraw 100 µl of the appropriate monomer solution (for the first coupling this is obviously the second base in the desired sequence since the 3'-terminal base is attached to the support) into the syringe and inject it into a vial of MSNT.

Phosphotriester Method

(xii) Leaving the syringe in place, agitate the vial gently to encourage dissolution of the MSNT and inject 10 μl of 1-methylimidazole into the vial after 1 min using a separate 50 μl gas-tight syringe. Be very careful not to damage the delicate tips of the syringe needles.

(xiii) After the pyridine wash in step 5, turn the solvent selector to the stop position, unscrew the septum cap of the column about half a turn and slowly insert the syringe containing activated monomer solution (total volume will be ~125 μl).

(xiv) Tighten the cap back half a turn and inject the monomer solution gently over about 15 sec.

(xv) Unscrew the cap half a turn again, withdraw the syringe carefully and retighten the screw cap. Cover it with Parafilm, and then close the tap on the column outlet.

(xvi) Rinse the syringe out thoroughly five or six times with clean pyridine and then leave full of dry pyridine to keep the glass barrel water-free until next required.

(xvii) When the 15 min coupling time has elapsed, open the column outlet and commence the next cycle.

(xviii) Carry out the requisite number of cycles and after the final coupling wash the support with pyridine (5 min) followed by dichloroethane (3 min).

(xix) Stop the flow, disconnect the column and push the support into a small sintered glass funnel using the unlocked bottom plunger.

(xx) Wash the support with a few ml of dichloroethane followed by a few ml of dry diethyl ether.

(xxi) Check the weight of the dry support and transfer to a small plastic Eppendorf tube ready for deprotection.

4.3.1 *Trouble Shooting*

(i) If a large vapour or gas bubble appears above the support during a cycle (small bubbles can be ignored) stop the flow after the pyridine wash and inject about half a syringe (500 μl size) full of dry pyridine into the column, then slowly pull up the syringe plunger and the bubble will be withdrawn into the syringe. Run pyridine for a few seconds to make sure that the bubble has been completely removed, then remove the syringe and continue the assembly.

(ii) Should the flow-rate suddenly drop and/or if the support drains then it is quite probable that a piece of rubber septum has become trapped in the bore of the column top piece below the solvent inlet. If this is the case remove the column top carefully, replace the septum and clear the bore of the fitting with a piece of fine wire to remove any obstruction. Then reassemble the column, remove any air as described above and continue as normal. Remember to cover the top of the column while cleaning out the top piece.

Table 5. Wash and Cycle Sequence for Assemblies on about 60 mg of Polyamide Support.

Step	Operation	Time (min)
1	Pyridine wash	5
2	1,2-dichloroethane wash	4
3	Deprotection with 10% trichloroacetic acid in 1,2-dichloroethane[a]	3
4	DMF wash	2
5	Pyridine wash	6
6	Coupling (stop flow)	15

[a]Prepare as follows: weigh 20 g of trichloroacetic acid (**Warning: corrosive and very hygroscopic**) into a 200 ml Omnifit bottle and dry thoroughly *in vacuo* over anhydrous calcium sulphate for 24 h. Release the vacuum with dry nitrogen, fill the bottle with 200 ml of dry 1,2-dichloroethane and immediately screw on the top. Swirl the contents gently to obtain solution.

4.4 Large-scale Synthesis of Oligodeoxyribonucleotides

Using the polydimethylacrylamide-Kieselguhr support you can easily make milligram quantities of oligodeoxyribonucleotides. For example, starting with 60 mg of the low loading support (80 – 90 µmol nucleoside per gram) you should be able to obtain about 3 – 4 mg of a purified 20-residue oligonucleotide, whereas using 150 mg of the high loading support (~180 – 200 µmol nucleoside per gram) you can expect about 20 mg of a purified 20-residue oligonucleotide.

To carry out an assembly on 60 mg (~5 µmol) of the low loading support use 40 mg of MSNT, 40 µmol of monomer and 35 µl of 1-methylimidazole in 350 µl of dry pyridine per coupling reaction. Prior to starting the first cycle (refer to Sections 4.2 and 4.3 for general details) treat the support with 0.5 ml of a 10% solution of phenyl isocyanate in dry pyridine for 15 min to dry it. Use the wash cycle shown in *Table 5*.

Use a flow-rate of about 1 ml per minute in pyridine. (N.B. after the acid wash, the support will still be orange coloured. However, this is quite normal, and the colour will disappear during the DMF wash.) If there is any uncertainty about whether the deprotection is complete wash the support with pyridine, dichloroethane and then repeat the acid flow (briefly) to see whether it turns orange again before proceeding. Under normal circumstances you will not need to do this. Note that the support swells and contracts with change of solvent, and its volume will increase considerably as the oligodeoxyribonucleotide chain extends so you will need to allow for this by occasional adjustment of the bottom plunger.

If you wish to work on a larger scale [e.g., 150 mg of high loading support (~30 µmol)], use about 250 µmol of monomers, 230 mg of MSNT and 150 µl of 1-methylimidazole in 1.5 ml of dry pyridine per coupling. Because of the large volume increase and subsequent alteration of flow-rates it is advisable to divide this into three columns containing 50 mg of support run in parallel. It is advisable to puncture the frit with about 10 needle holes so that any fines generated do not block it (if using >50 mg in a column it is best to cut the frit in half to reduce its thickness). Failure to do this can result in greatly reduced solvent flow or even

Phosphotriester Method

blockage. Since the swollen support will occupy a substantial volume and because it is best not to exceed about 6 p.s.i. of gas pressure (to avoid support compression), increase the various wash times such that about 10 bed volumes pass through the support during each wash (use a measuring cylinder to monitor this). Failure to do this may result in incomplete removal of TCA causing a substantial capping reaction of the 5′-hydroxyl group during the condensation step.

5. CLEAVAGE OF THE OLIGODEOXYRIBONUCLEOTIDES FROM THE SUPPORT AND SUBSEQUENT DEPROTECTION

(i) Place the dry support bearing the fully protected oligodeoxyribonucleotide prepared according to Section 4.3, in a 2 ml plastic Eppendorf tube.

(ii) Prepare a solution of either *syn*-2-nitrobenzaldoxime (70 mg, 0.422 mmol) or pyridine-2-carbaldoxime (51.5 mg, 0.422 mmol) in dioxan/water (1 ml, 1:1 v/v) and add 1,1,3,3-tetramethylguanidine (50 μl, 0.4 mmol).

(iii) Add the required volume of this oximate solution to the Eppendorf tube containing the support, cap the tube and seal well with Parafilm or Nescofilm. To determine the amount of oximate required multiply the amount of 3′ starting nucleoside by the number of phosphotriester moieties and multiply that by 10 to give a reasonable excess; for example, use 0.5 ml of the solution to deprotect a 21-residue oligonucleotide prepared on a 1 μmol scale (note that the nitrobenzaldoximate solution is orange coloured whereas the pyridine-2-carbaldoximate solution is very pale yellow).

(iv) Shake the Eppendorf tube briefly and leave to stand for about 16–20 h at room temperature or overnight at 37°C. Besides cleaving the 2-chlorophenyl protecting groups and the succinate linkage, the oximate reagent reverses some base modifications, particularly at O^6 of dG, so for G-rich and very long sequences use an extended treatment (30–40 h at room temperature or 20 h at 37°C) since the reversal is rather slow.

(v) Next, filter the mixture through a small sintered glass funnel into a 50 ml round-bottomed flask and wash the support with dioxan/water (1:1 v/v) until the washings are colourless.

(vi) Evaporate the filtrate to dryness *in vacuo* and dissolve the residue in concentrated aqueous ammonia (5 ml, specific gravity 0.88).

(vii) Seal the flask well (use a greased stopper and plenty of tape) or use a sealed vial and heat at 60°C for 15–20 h. (N.B. if N^6-benzoyl protection on dA has been used then this time can be reduced to about 6 h.) It is essential that the ammonia does not leak out substantially otherwise deprotection may be incomplete.

(viii) When cool, open the flask carefully and remove the grease from the joint with chloroform or ether, then carefully evaporate the solution to dryness *in vacuo*.

(ix) Dissolve the residue in acetic acid/water (5 ml, 8:2 v/v) and leave the solution at room temperature for 30 min to effect loss of the 5′-O-pixyl or DMTr group.

(x) Add water (5 ml) and extract with diethyl ether (10 ml) five times in a separating funnel, shaking vigorously each time.

(xi) Evaporate the aqueous layer (bottom one) to dryness *in vacuo* (keep the bath temperature below 40°C).
(xii) Finally, co-evaporate with water (~5 ml) twice to remove any residual acetic acid. Prior to the last evaporation filter the solution through a 0.45 μm disposable filter to remove any particulate matter.
(xiii) Dissolve the final residue in water (0.5 ml, de-ionised) ready for purification.

6. PURIFICATION PROCEDURES

6.1 Ion-exchange H.p.l.c.

Purify the crude oligodeoxynucleotide obtained from Section 5 on a Partisil 10SAX column (or equivalent ion-exchange column) using a linear concentration gradient from 1 mM to 0.3 M KH_2PO_4, pH 6.3, in formamide/water (6:4 v/v). A purine-rich oligonucleotide will be retained more strongly than a pyrimidine-rich one of the same length. The formamide prevents aggregation phenomena which can be a nuisance with G-rich sequences. If the sequence is either self-complementary, as will be the case with a restriction site linker, or if the sequence can form a stable hairpin structure it will be necessary to thermostat the column at about 50°C in order to obtain narrow h.p.l.c. peaks. Refer to Chapter 5 for full details on chromatographic purification of synthetic oligonucleotides. Oligonucleotides up to about 40 bases in length can be purified by this method if an extended gradient to 0.4 M KH_2PO_4 is used.

As an alternative and indeed the only way to purify very long oligodeoxyribonucleotides, use polyacrylamide gel electrophoresis, details of which are to be found in Chapter 6.

6.2 Desalting

Remove the salt and formamide from the sample of ion-exchange h.p.l.c. purified material either by passage through a column of Bio-Gel P-2 (use a column volume at least 10 times that of the volume of solution to be desalted) eluted with ethanol/water (2:8 v/v) or by dialysis against water. Use Visking or Spectropor (low molecular weight cut off) tubing which has been boiled in 2% sodium bicarbonate, 1 mM EDTA and then boiled in distilled water. (This method cannot be used for oligodeoxyribonucleotides shorter than 6–7 residues.)

The desired material will elute at the void volume of the column. Evaporate those fractions containing the product to dryness *in vacuo* and dissolve the residue in an appropriate volume of water and measure the absorbance of the solution at 260 nm on a spectrophotometer. Calculate the yield after ion-exchange h.p.l.c. and desalting as follows:

Overall yield (Y_o) =

$$\frac{A_{260} \text{ units obtained} \times 100}{\text{Fraction of material purified} \times \Sigma\epsilon_{260} \text{ of each base} \times \mu\text{mol of starting resin}} \%$$

where ϵ_{260} is 8.8 for T; 7.3 for C; 11.7 for G; and 15.4 for A in $cm^2/\mu mol$.

Phosphotriester Method

The repetitive or coupling yield (Y_c) is related to the overall yield (Y_o) as follows:

$$Y_c = \left(\frac{Y_o}{100}\right)^{1/n} \%$$

where n is the number of coupling cycles. In practice Y_c is in the range of 88–92% depending upon the support used and the base composition of the oligodeoxyribonucleotide. (N.B. no allowance has been made for hypochromicity so the actual yield may be higher than that calculated as above.)

6.3 Reversed Phase H.p.l.c.

This is normally used as an analytical check on the purity of the oligodeoxyribonucleotide purified by ion-exchange h.p.l.c. Generally the ion-exchange purified materials are sufficiently pure for most purposes, particularly if they are to be used in cloning work. However, very pure oligodeoxyribonucleotides are required for structural studies, X-ray crystallography and n.m.r. spectroscopy work, so preparative reversed phase h.p.l.c. on a μ-Bondapak C_{18} column (or equivalent) is needed as the final purification step. Refer to Chapter 5 for full details of reversed phase h.p.l.c., including recommended solvent systems.

After the final purification it is good practice to check the sequence of the oligodeoxyribonucleotide according to the methods described in Chapter 6.

7. CHEMICALS AND EQUIPMENT

7.1 Supports

	Item	Supplier	Catalogue No.
(i)	Unfunctionalised polydimethylacrylamide-Kieselguhr resin (Wolfchem SR 108)	Cruachem	7090
	Fmoc-gly-gly-resin	Cruachem	7000
	DMTrdT succinyl-polyamide resin	Cruachem	7200
	DMTrdibG succinyl-polyamide resin	Cruachem	7210
	DMTrdbzA succinyl-polyamide resin	Cruachem	7220
	DMTrdbzC succinyl-polyamide resin	Cruachem	7230
(ii)	Long chain alkylamine controlled pore glass (LCAA/CPG) (500 Å pore diameter; 125–177 μm particle size)	Pierce	24875
	DMTrdT succinyl-LCAA/CPG	Cruachem	7620
	DMTrdibG succinyl-LCAA/CPG	Cruachem	7621
	DMTrdbzA succinyl-LCCA/CPG	Cruachem	7622
	DMTrdbzC succinyl-LCAA/CPG	Cruachem	7623

7.2 Succinates

DMTrdT-3'-O-succinate	Cruachem	7400
DMTrdibG-3'-O-succinate	Cruachem	7410
DMTrdbzA-3'-O-succinate	Cruachem	7420
DMTrdbzC-3'-O-succinate	Cruachem	7430

7.3 2'-Deoxyribonucleosides

Thymidine	Cruachem	1900
2'-Deoxyguanosine	Calbiochem	2600
	Cruachem	1910
2'-Deoxyadenosine	Calbiochem	2560
	Cruachem	1920
2'-Deoxycytidine hydrochloride	Calbiochem	2580
	Cruachem	1930

7.4 5'-O-Pixyl Protected 2'-Deoxynucleosides

PxdT	Cruachem	2200
PxdibG	Cruachem	2210
PxdbzA	Cruachem	2220
PxdbzC	Cruachem	2230

7.5 5'-O-Dimethoxytrityl Protected Monomers

DMTrdT-3'-O-(2-chlorophenyl phosphate)	Cruachem	7300
DMTrdibG-3'-O-(2-chlorophenyl phosphate)	Cruachem	7310
DMTrdbzA-3'-O-(2-chlorophenyl phosphate)	Cruachem	7320
DMTrdbzC-3'-O-(2-chlorophenyl phosphate)	Cruachem	7330

7.6 Solvents

Any reputable supplier can be used. However, wherever possible obtain analytical grade solvents. Those required are acetone; benzene; chloroform; 1,2-dichloroethane; dichloromethane; diethyl ether; N,N-dimethylformamide (DMF); 1,4-dioxan; ethanol; ethyl acetate; formamide; methanol; n-pentane; pyridine; tetrahydrofuran (THF) and toluene.

7.7 Miscellaneous Reagents (with suggested suppliers)

Item	Supplier	Catalogue No.
Acetic anhydride (analytical grade)	BDH	10002
3-Amino-1,2,4-triazole	Fluka	09530
Acetic acid (analytical grade)	BDH	45001
Barium oxide	BDH	27952
Basic alumina	Woelm Pharma	04568
Benzophenone	BDH	27338
Benzoyl chloride (analytical grade)	BDH	10054
Bio-Gel P-2	Bio-Rad	150-0150
Calcium chloride (granular)	BDH	27587
2-Chlorophenyl phosphorodichloridate	Fluka	25928
Concentrated aqueous ammonia	BDH	10012
1,2-Diaminoethane (ethylene diamine)	BDH	28016
Dichloroacetic acid (puriss.)	Fluka	35810
N,N'-Dicyclohexylcarbodiimide (DCCI)	Fluka	36650
N,N-Diisiopropylethylamine	Fluka	03440
4-Dimethylaminopyridine	Aldrich	10,770-0
N,N-Dimethylformamide dimethyl acetal	Aldrich	14,073-2
Di-n-butylamine	Aldrich	D4,495-2
Dowex 50W-X8 resin, pyridinium form	Cruachem	1500
9-Fluorenylmethoxycarbonylglycine (Fmoc-glycine)	Fluka	47627
Isobutyric anhydride	Fluka	58390
Kieselgel 60H	Merck	7736
Mesitylenesulphonyl chloride (puriss.)	Fluka	63932
MSNT	Cruachem	1085
1-Methylimidazole	Fluka	67560
Ninhydrin (analytical grade)	BDH	10132
syn-2-Nitrobenzaldoxime	Cruachem	1210
3-Nitro-1,2,4-triazole	Cruachem	1112
Perchloric acid (60% aqueous, analytical grade)	BDH	10175
Phenyl isocyanate (puriss.)	Fluka	78750
Piperidine	BDH	29565
Phosphorus pentoxide	BDH	29527
Pixyl chloride	Fluka	26035

Potassium dihydrogen phosphate (AnalaR)	BDH	10203
Potassium hydroxide pellets	BDH	10210
Pyridine-2-carbaldoxime	Fluka	82760
Silica gel (self-indicating)	BDH	30062
Sodium metal	BDH	30101
Sodium hydrogen carbonate	BDH	10247
Sodium nitrite (analytical grade)	BDH	10256
Sodium sulphate (anhydrous, analytical grade)	BDH	10264
Succinic anhydride (puriss.)	Fluka	14089
1,1,3,3-Tetramethylguanidine (puriss.)	Fluka	87844
1,2,4-Triazole (puriss.)	Fluka	90630
Trichloroacetic acid (analytical grade)	BDH	10286
Triethylamine (analytical grade)	BDH	10409
Trimethylchlorosilane	Fluka	92362

7.8 Special Equipment

	Item	Supplier	Catalogue No.
(i)	5 ml pear-shaped flasks (B14 neck)	Quickfit	
	Silicone rubber Suba-seal stoppers (size no. 30) to fit the above	Gallenkamp	SYJ 350 200T
(ii)	(Alternative to (i) above)		
	Box of 12 x 3 ml Reacti-Vials	Pierce	13222
	or 12 x 5 ml Reacti-Vials	Pierce	13223
	Package of 6 dozen Tuf-Bond Teflon/Silicone discs (18 mm diameter to fit 3 ml and 5 ml Reacti-Vials)	Pierce	12718
(ii)	Several boxes of 12 x 1 ml microvials	Pierce (or Camlab)	13221 MV10
	Package of 6 dozen Tuf-Bond discs (12 mm diameter)	Pierce	12712
(iv)	Gas-tight syringes:		
	1725RN (250 μl)	Hamilton	81130
	1750RN (500 μl)	Hamilton	81230
	1001RN (1 ml)	Hamilton	81330
	RN needle package (pack of 3, 22 gauge, 2 in) to fit above syringes	Hamilton	80725
	50 μl gas-tight syringe (fixed needle) 1705N	Hamilton	

7.9 T.l.c. Plates

T.l.c. aluminium sheets coated with 0.2 mm silica gel 60F$_{254}$ (20 x 20 cm) — Merck 5554

7.10 Parts List for Dual Column Omnifit Synthesiser

1 x Pressure stat (50 p.s.i.)

4 x 1 litre 3-valve inert reservoir

1 x 100 or 200 ml 3-valve inert reservoir

2 x 50 mm column

2 x Septum injector end piece

2 x Variable length end piece

1 x Mounting for valve and column (includes clips, Rheodyne 6-way valve, and 2-way valve)

1 x 3-way valve

1 x 25 x 150 mm complete column (for drying tube)

1 x 8-way connector

4 x Brass pinch clamp

1 x 2-way polypropylene coupling

2 m of 1.5 mm i.d. Teflon tubing

2 m of 0.8 mm i.d. Teflon tubing

2 m of 0.3 mm i.d. Teflon tubing

Pack of Teflon faced silicone septa (Catalogue No. 3302)

Pack of Teflon frits (Catalogue No. 6652)

Assorted packs of plugs, tube end fittings, gripper fittings, silicone O-rings, and fluoron O-rings

8. REFERENCES

1. Miyoshi,K. and Itakura,K. (1980) *Nucleic Acids Research Symposium Series,* No. **7**, IRL Press, Oxford and Washington, D.C., p. 281.
2. Miyoshi,K. and Itakura,K. (1979) *Tetrahedron Lett.,* **19**, 3635 [and (1980) *Nucleic Acids Res.,* **8**, 5491].
3. Gait,M.J., Singh,M., Sheppard,R.C., Edge,M.D., Greene,A.R., Heathcliffe,G.R., Atkinson,T.C., Newton,C.R. and Markham,A.F. (1980) *Nucleic Acids Res.,* **8**, 1081.
4. Gait,M.J., Popov,S.G., Singh,M. and Titmas,R.C. (1980) *Nucleic Acids Research Symposium Series,* No. 7, IRL Press, Oxford and Washington, D.C., p. 243.
5. Duckworth,M.L., Gait,M.J., Goelet,P., Hong,G.F., Singh,M. and Titmas, R.C. (1981) *Nucleic Acids Res.,* **9**, 1691.
6. Gait,M.J., Matthes,H.W.D., Singh,M., Sproat,B.S. and Titmas,R.C. (1982) *Nucleic Acids Res.,* **10**, 6243.
7. Kohli,V., Balland,A., Wintzerith,M., Sauerwald,R., Staub,A. and Lecocq, J.P. (1982) *Nucleic Acids Res.,* **10**, 7439.
8. Ohtsuka,E., Takashima,H. and Ikehara,M. (1982) *Tetrahedron Lett.,* **23**, 3081.
9. Gough,G.R., Brunden,M.J. and Gilham,P.T. (1981) *Tetrahedron Lett.,* **22**, 4177.
10. Köster,H., Stumpe,A. and Wolter,A. (1983) *Tetrahedron Lett.,* **24**, 747.
11. Sproat,B.S. and Bannwarth,W. (1983) *Tetrahedron Lett.,* **24**, 5771.

12. Frank,R., Heikens,W., Heisterberg-Moutsis,G. and Blöcker,H. (1983) *Nucleic Acids Res.*, **11**, 4365.
13. Matthes,H.W.D., Zenke,W.M., Grundström,T., Staub,A., Wintzerith,M. and Chambon,P. (1984) *EMBO J.*, **3**, 801.
14. Adams,S.P., Kavka,K.S., Wykes,E.J., Holder,S.B. and Gallupi,G.R. (1983) *J. Am. Chem. Soc.*, **105**, 661.
15. Atherton,E., Brown,E., Sheppard,R.C. and Rosevear,A. (1981) *J. Chem. Soc. Chem. Commun.*, 1151.
16. Reese,C.B. (1970) *Colloques Internationaux du CNRS*, **182**, 319.
17. Silber,G., Flockerzi,D., Varma,R.S., Charubala,R., Uhlmann,E. and Pfleiderer,W. (1981) *Helv. Chim. Acta*, **64**, 1704.
18. Reese,C.B. and Zard,L. (1981) *Nucleic Acids Res.*, **9**, 4611.
19. Chattopadhyaya,J.B. and Reese,C.B. (1978) *J. Chem. Soc. Chem. Commun.*, 639.
20. Kume,A., Sekine,M. and Hata,T. (1982) *Tetrahedon Lett.*, **23**, 4365.
21. Kume,A., Sekine,M. and Hata,T. (1983) *Chem. Lett.*, **1597**.
22. Froehler,B.C. and Matteucci,M.D. (1983) *Nucleic Acids Res.*, **11**, 8031.
23. Reese,C.B. and Ubasawa,A. (1980) *Nucleic Acids Research Symposium Series*, No. 7, IRL Press, Oxford and Washington, D.C., p. 5 [and (1980) *Tetrahedron Lett.*, **21**, 2265].

UPDATE

In place of the succinate linkage (section 2.2), a new linkage has been introduced by us (24) to join a deoxyribonucleoside to controlled pore glass. Briefly, an appropriately protected 2′-deoxyribonucleoside bearing a free 3′-hydroxyl group is reacted with an equivalent of tolylene-2,6-diisocyanate (Aldrich) in the presence of N,N-diisopropylethylamine to generate an intermediate monoisocynate. Without isolation, this is reacted further with an equivalent of LCAA/CPG to obtain the fully functionalised support (24 – 34 µmol per gram of deoxyribonucleoside). Use of the new linkage obviates the need for preparation of deoxyribonucleoside-3′-O-succinates. Moreover, the greater stability of the urethane moiety in the new linkage, compared to the ester moiety in the succinate linkage, results in more reliable syntheses, since the new linkage is not susceptible to cleavage by basic impurities in the solvents used in chain assembly. The linkage is cleaved at the end of the assembly by treatment with concentrated ammonia solution. Full details are given in reference 24.

The method of simultaneous synthesis of oligodeoxyribonucleotides (also known as 'segmented' synthesis using cellulose paper discs (12,13) (section 2.2) has received further support in two recent publications, one advocating phosphotriester chemistry but with slightly altered conditions of 5′-terminal and final deprotection (25), the other describing the use of phosphoramidite chemistry (26). Whilst expermentation with these new techniques is to be encouraged, variability of results obtained between batches of paper, even individual sheets (see Cruachem Highlights, March 1985 edition), means that the method is not yet to be recommended to the complete novice, or for the synthesis of long chains. However, it will clearly not be long before these minor problems are remedied, such that the full potential of this exciting technique can be recognised.

Finally the use of 4-morpholino pyridine-N-oxide instead of N-methylimidazole in coupling reactions leads to a reduction in the time of coupling to 2 – 5 minutes (27).

25. Sproat,B.S. and Brown,D.M. (1985) *Nucleic Acids Res.*, **13**, 2979.
26. Brenner,D.G. and Shaw,W.V. (1985) *EMBO J.*, **4**, 561.
27. Efimov,V.A., Chakhmakhcheva,O.G. and Ovchinnikov,Yu.A. (1985) *Nucleic Acids Res.*, **13**, 3651.

CHAPTER 5

Chromatographic Purification of Synthetic Oligonucleotides

LARRY W. McLAUGHLIN AND NORBERT PIEL

1. INTRODUCTION

Chromatography involves the partition of a solute between a mobile phase and a stationary phase. An oligonucleotide dissolved in the mobile phase and introduced into the inlet of a column containing the stationary phase will migrate through the column at a rate dependent on its interaction between these two phases. Since oligonucleotides are essentially polyanions containing the lipophilic nucleobases, anion-exchange and reversed phase chromatography are generally useful for isolation and analysis of a particular product.

Anion-exchange chromatography depends upon the adsorption and desorption of the anionic solute on a cationic stationary phase. These are equilibrium processes which depend upon the nature of the eluting buffer. For a given set of conditions, desorption of a mixture of polyanionic solutes will depend upon the net charge of each polymer or, in the case of oligonucleotides, the chain length.

Traditional adsorption chromatography is performed on a polar stationary phase such as silica gel and uses a non-polar mobile phase. In this 'normal phase' chromatography polar solutes are tightly bound to the polar stationary phase and are thus eluted later than non-polar solutes. Polar compounds are eluted from the stationary phase by increasing the polarity of the mobile phase. In reversed phase chromatography, as implied, the conditions are reversed. The stationary phase is non-polar and the mobile phase is polar. It is the hydrophobic interactions which determine the velocity of migration along the stationary phase. Polar solutes are eluted relatively early from the column. Elution of a particular compound in this case is effected by reducing the polarity of the aqueous mobile phase usually by the addition of an organic solvent.

Recently a third type of stationary phase has been developed for the resolution of oligonucleotide and nucleic acid mixtures using both ionic and hydrophobic interactions (see Section 5).

2. EQUIPMENT FOR HIGH PERFORMANCE LIQUID CHROMATOGRAPHY

With the increased interest in the application of h.p.l.c. to the analysis and isolation of a wide variety of chemical and biochemical products, a number of commercially produced h.p.l.c. systems have become available. While it is very easy (although potentially expensive) to purchase a complete system (including test samples) it may be valuable to consider how an h.p.l.c. system functions and where costs can be saved.

The chromatographic interactions which occur during analysis by h.p.l.c. are no different from those found in other chromatographic systems. The only difference one encounters with h.p.l.c. is that column inlet pressures are often in excess of 100 bar (1420 p.s.i.), hence h.p.l.c. has sometimes been called high pressure liquid chromatography. In order to perform liquid chromatography accurately and reproducibly at these increased pressures it has been necessary to develop associated technologies.

A basic h.p.l.c. system includes one or two pumps with associated gradient former, an injection port, column, detector and recorder. Many of these components are available in scientific laboratories and can, in some cases, be adapted for the construction of an h.p.l.c. system. One difficulty in building an h.p.l.c. system is versatile and reproducible gradient formation. There are generally two possibilities. In the first method the gradient is mixed on the high pressure side of the pump and two pumps are required, one for buffer A and one for buffer B. The gradient is formed by controlling the individual pumping velocities using a microprocessor. A high pressure mixing chamber lies between the pumps and the column. In the second method the gradient is mixed on the low pressure side of the pump. Two proportioning values controlled by a microprocessor determine the amount of either buffer which is admitted to a low pressure mixing chamber. Following the mixing chamber is the inlet valve to a single pump. Both systems result in reproducible gradient formation.

In order to simplify the introduction of the desired sample to the inlet of the column a number of injection port designs are available which allow injection of a sample at atmospheric pressure into an isolated sample loop. This sample loop is then connected through a series of high pressure seals to the inlet of the column. Probably the most widely used and trouble-free injection system available is the Rheodyne 7120 or 7125 injection port. This can generally be purchased less expensively directly from Rheodyne than from the various companies producing h.p.l.c. equipment.

A column oven is desirable since enhanced resolution is often obtained at temperatures above ambient temperature. A number of ovens are available. A less expensive possibility is to simply prepare a large diameter tube (plastic or metal) in which the column can be sealed. With inlet and outlet tubing, water of a desired temperature can be pumped around the column using a temperature-controlled water-circulating bath, as available in most laboratories.

A detector is necessary in order to monitor the column effluent. Several possibilities are available, but owing to the high extinction coefficients exhibited by nucleosides, nucleotides and oligonucleotides in the range 250–280 nm, u.v. detectors are almost exclusively used. Two types of u.v. detectors are generally available, filter photometric detectors and spectrophotometer detectors. In the former, the wavelength monitored is determined from the lamp and filter used and a number of possibilities are usually available. In the latter, the column effluent can be monitored at any specific wavelength from 200 to 320 nm. In some cases it is also possible to purchase a preparative flow-through sample cell. This usually reduces detector sensitivity by a factor of 10, which may be necessary for preparative isolations. On the other hand it is also possible to monitor the column effluent at a wavelength where the extinction coefficient is not so high (280−300 nm) during preparative runs.

Almost any laboratory recorder can be used to record the detector output in an h.p.l.c. system; purchasing a new recorder is generally not necessary.

3. THE CHROMATOGRAPHIC SUPPORT

The column, and more specifically its stationary phase, is the heart of any liquid chromatographic system and will ultimately determine the resolution of a given oligonucleotide mixture. The majority of commercially available chromatographic matrices contain porous, spherical or irregularly shaped pressure-stable particles to which the ion-exchange or reversed phase stationary phase has been chemically bound.

3.1 Bonded Phase Silica Supports

Most h.p.l.c. column supports available today consist of bonded phase, microparticulate (5−30 μm) silica gels. These are prepared by reacting porous silica gel particles with an organic silyl chloride. The organic moiety can vary considerably. This allows some control over the polarity of the stationary phase. Extremely hydrophobic matrices are prepared by attaching octadecylsilyl (C_{18}) residues to the silica gel. Variations in the hydrophobicity are obtained by using shorter C_8, C_6, C_4, methyl or phenyl residues. More polar stationary phases can be produced by binding cyanopropyl, nitropropyl or aminopropyl residues to the silica gel backbone. Anionic or cationic resins can be prepared by binding quaternary amines or sulphonic acids respectively to the porous silica beads.

Bonded phase silica supports from different suppliers will vary considerably in shape (spherical or irregular), carbon loading (amount of bound organic moiety), extent of monolayer coverage, pore size and whether or not the silica is capped after modification. After a number of trial runs we have found exceptional resolution of oligonucleotides and their constituents using bonded phase supports prepared from Hypersil (a product of Shandon, Southern Co, Runcorn, UK). Hypersil consists of spherical particles (3 μm, 5 μm or 10 μm) with relatively large pores (110 nm average diameter) and low carbon loading (10%) implying a high degree of monolayer coverage. Additionally the support is capped, such that there should be no influence of free silanol (Si-OH) groups on the separation mechanism. Bonded phase, Hypersil supports result in columns of very high quality when packed as described in Appendix II.

3.2 Pre-packed or Self-packed Columns

After the initial investment in an h.p.l.c. system, the columns, which must be regularly renewed, are generally the single most expensive item involved in the subsequent running costs. This cost can be vastly decreased by learning to prepare self-packed columns. It is generally considered that commercially pre-packed columns will always have a higher theoretical plate count than self-packed columns.

Whether or not this is true is questionable, nevertheless the retort is that columns should be packed to perform separations not to count theoretical plates.

If the economics of pre-packed *versus* self-packed columns are considered it is immediately obvious that money can be saved by purchasing chromatographic supports in bulk and columns for self-packing. A simple procedure to pack analytical columns of high quality is described in Appendix II.

There is not generally as large a price difference between the cost of pre-packed preparative columns and the cost of the required bulk packing material and empty column. In this case it may be more efficient to purchase a column containing the desired stationary phase. However, subsequent repacking of the larger columns will in time also result in significant cost savings (see Appendix II).

4. SELECTED APPLICATIONS

This section describes the isolation and analysis of chemically synthesised oligodeoxyribonucleotides, the isolation and analysis of chemically synthesised oligoribonucleotides and the isolation and analysis of oligoribonucleotides resulting from enzymatic syntheses. Two types of chromatography (anion-exchange and reversed phase) are discussed with respect to isolation and analysis of a particular product. In anion-exchange chromatography resolution is related to the number of phosphate groups present. In reversed phase chromatography interactions with the nucleobases primarily determine the extent of separation.

4.1 Separation of Synthetic Oligodeoxyribonucleotide Mixtures

The chemical synthesis of oligodeoxyribonucleotides on solid-phase supports is well established (see Chapters 3 and 4). Isolation methods are described here for resolution of oligodeoxyribonucleotide mixtures which result from such synthetic procedures. While the example chromatograms are taken from oligodeoxyribonucleotide mixtures which have been synthesised on solid-phase supports using 3'-phosphorylated deoxyribonucleotides which have been activated with 1-hydroxybenzotriazole (1), the methods are also applicable to mixtures resulting from similar phosphotriester or phosphite-triester synthetic procedures.

After completion of the chemical synthesis of an oligodeoxyribonucleotide it is generally desirable to analyse the synthesis mixture to determine if a significant amount of desired oligomer is present. This initial analysis could in theory by accomplished by either anion-exchange or reversed phase chromatography. However, the desired oligomer, by virtue of being the longest oligomer, carries the largest number of phosphate groups. Analysis by anion-exchange chromatography will produce a chromatogram in which the product elutes as the last peak of the oligonucleotide mixture. This is not the case in reversed phase chromatography where the chromatographic interactions are primarily with the nucleobases. Since purine-rich sequences are more strongly retained than pyrimidine-rich sequences on the hydrophobic stationary matrix, the desired oligomer often elutes in the middle of the chromatogram which complicates identification and subsequent calculation of the yield of a synthesis.

(i) Prior to chromatographic analysis, remove the phosphate protecting groups and the nucleobase protecting groups using the procedures described in Chapters 3 and 4.
(ii) Dissolve the residue in 2-3 ml water and extract three times with an equivalent amount of diethyl ether. This removes the insoluble material resulting from the deprotection reactions as well as a large portion of the soluble material, which generally elutes at the void volume in the subsequent chromatographic analysis.
(iii) Evaporate a small aliquot of this mixture (10 μl) to dryness and treat with 80% acetic acid/water at 0°C for 30 min.
(iv) Evaporate the acidic mixture once again and then evaporate the residue three times from water. This treatment removes the 5'-terminal 4,4'-dimethoxytrityl or 9-phenylxanthenyl group prior to analysis by anion-exchange chromatography.
(v) Dissolve the remaining residue in 10-30 μl of distilled water.

The analysis can be performed on two types of stationary phases. The first contain quaternary ammonium ions bound through a short carbon linker to the silica backbone of the support and are generally known as strong anion-exchangers (SAX). The second contain a primary, secondary or tertiary amine bound also through a short carbon linker to the silica backbone. These latter supports will act as anion-exchangers in acidic buffers in which case the amine groups are protonated. Since the ability of these latter supports to act as anion-exchange resins is related to the number of protonated amine groups, retention times on these stationary phases will be strongly dependent on the pH of the mobile phase.

Figure 1 illustrates the analysis of the crude oligodeoxyribonucleotide mixture resulting from the synthesis of the decamer d(GpGpGpGpApApApCpCpT) on a silica gel support. The formamide buffer described in the legend to *Figure 1* in

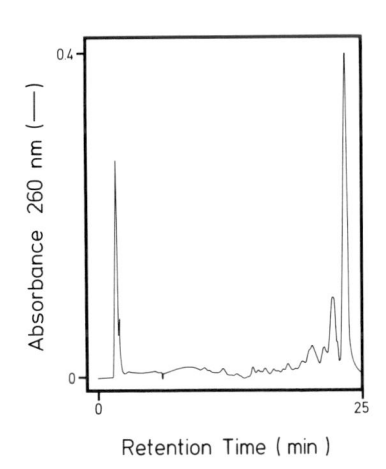

Figure 1. Analysis of the crude mixture containing the decamer d(GpGpGpGpApApApCpCpT) on a 4.6 x 250 mm column of Whatman SAX (10 μm): buffer A: 0.001 M KH$_2$PO$_4$ (pH 6.3), 60% formamide; buffer B: 0.3 M KH$_2$PO$_4$ (pH 6.3), 60% formamide; gradient: 0-75% buffer B in 45 min; flow 2.0 ml/min; temperature 45°C.

Chromatographic Purification of Synthetic Oligonucleotides

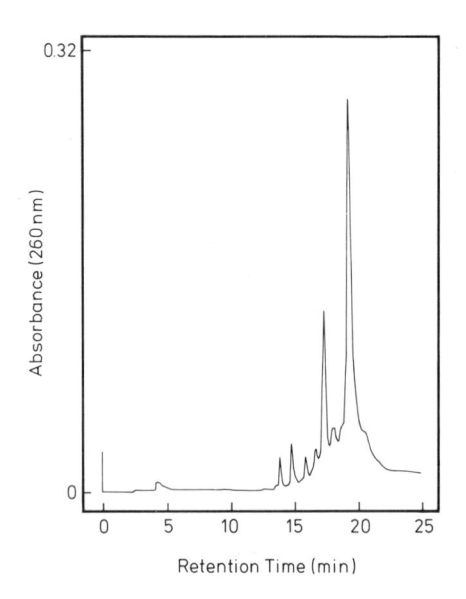

Figure 2. Analysis of the crude mixture containing d(GpApGpTpGpApGpCpTpApApCpTpCpApCpApT) on a 4.6 x 250 mm column of APS-Hypersil (5 μm): buffer A: 0.02 M KH_2PO_4 (pH 6.8); buffer B: 0.9 M KH_2PO_4 (pH 6.8); gradient 0–100% buffer B in 60 min; flow 1.5 ml/min; temperature 35°C.

combination with an SAX column as previously described (2) is recommended for oligodeoxyribonucleotide products which contain a number of deoxyguanosine residues where aggregation may result in poor resolution. It also inhibits base-pairing of oligomers which are self-complementary. In both cases the formamide-containing mobile phase will disrupt secondary structure effects and give cleaner analyses. The analysis is monitored at 270 or 280 nm. This is necessary to suppress absorption from u.v.-absorbing contaminants in the formamide, which are very intense at 260 nm. It is also important with this mobile phase that the formamide content of each buffer is carefully measured. If the two buffers have slightly different concentrations of formamide, a small formamide gradient will occur during elution which will be expressed as a drift of the baseline. The system described in *Figure 1* can of course be used for other oligonucleotide products. However, in our experience, the SAX supports are relatively unstable. With each analysis the retention times become significantly shorter until nothing is retained on the column. This process can in some cases result in the loss of a column in a matter of days.

Therefore, we have used, in addition to the SAX stationary phase, an aminopropylsilyl bonded phase (APS) for analysis of oligodeoxyribonucleotide mixtures. In *Figure 2* the analysis of the crude mixture containing the octadecamer p(GpApGpTpGpApGpCpTpApApCpTpCpApCpApT) is illustrated. The octadecamer was prepared on a controlled pore glass (CPG) bead support. In order that the oligomer could be eluted from the column within a reasonable time, a mobile phase of pH 6.8 was used. At lower pHs longer retention times will be observed. At a pH of 4.5 the octadecamer is not eluted from the column and remains irreversibly bound. Resolution of oligonucleotide mixtures on the APS column is in many cases comparable with that achieved on the SAX column. However, the APS column in our hands has a significantly longer lifetime than the SAX column. The mobile phase described in *Figure 2* does not resolve well oligomers containing a number of deoxyguanosine residues or self-complementary oligodeoxyribonucleotides. In these cases the formamide-containing mobile phase and the SAX stationary phase appear to give the best analytical results.

We now use the SAX column only for analyses of oligomers where secondary structural effects may be present, as exemplified by the analysis shown in *Figure 1*. In this case the decamer starts with four dG residues. In all other analyses by anion-exchange chromatography we use the APS column. This approach extends the life of a SAX column and tends to help reduce the overall costs of h.p.l.c.

4.2 Isolation of Oligodeoxyribonucleotide Products

Either the SAX column or APS column can be used additionally to isolate the desired oligodeoxyribonucleotide product. However, a subsequent desalting step is required to remove formamide and/or phosphate salts. We have also observed that the yield of oligodeoxyribonucleotide isolated by anion-exchange chromatography using either of these stationary phases tends to be quite low. Products isolated using hydrophobic stationary phases (reversed phase columns) are generally obtained in higher yields. The use of a reversed phase column, commonly octadecylsilyl (ODS) bound silica gel, and a volatile mobile phase (triethylammonium acetate) is often an easier approach to the isolation problem.

If during the procedures of chemical synthesis a 'capping' step is used after each coupling step then, at the completion of the synthesis, if there are no other side reactions, only the desired oligomer should carry a 5'-terminal 4,4'-dimethoxytrityl or 9-phenylxanthenyl protecting group. The failure sequences contain free 5'-hydroxyl groups which should have been acylated during the 'capping' step and then deacylated during final deprotection. Both of these protecting groups are highly lipophilic and will interact strongly with the ODS stationary phase. An oligodeoxyribonucleotide carrying either of these terminal protecting groups will have a longer retention time than the corresponding oligodeoxyribonucleotide with a free 5'-hydroxyl (3).

The use of a triethylammonium acetate buffer system in combination with the hydrophobic ODS column has two advantages. As noted above, the buffer is volatile and a subsequent desalting step is unnecessary. Secondly, the triethylammonium ions will pair with the negatively charged phosphate groups. This ion-pairing phenomenon will tend to order the elution of oligodeoxyribonucleotides according to the number of phosphate groups present. Using this buffer system it is very unlikely that a short oligodeoxyribonucleotide of high purine content would co-elute with the desired oligodeoxyribonucleotide. The desired product will be shifted to a longer retention time than the shorter failure sequences as a result of a larger number of phosphate groups and the lipophilic 4,4'-dimethoxytrityl or 9-phenylxanthenyl terminal protecting group.

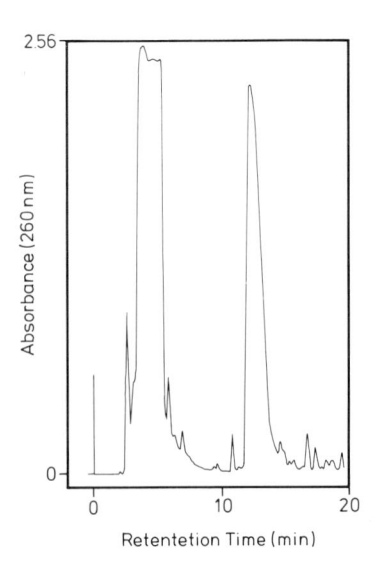

Figure 3. Isolation of the partially protected octamer 9-phenylxanthenyl-d(CpGpTpTpTpTpGpC) using a 9.4 x 250 mm column of ODS-Hypersil (5 μm): buffer A: 0.1 M ammonium acetate (pH 7.0); buffer B: 0.1 M ammonium acetate (pH 7.0), 80% acetonitrile; gradient: 25–65% buffer B in 40 min; flow 4.0 ml/min, temperature 35°C.

4.2.1 Isolation of an Octadeoxyribonucleotide

Isolation of the octamer 9-phenylxanthenyl-d(CpGpTpTpTpTpGpC), prepared on a Kieselgur-polydimethylacrylamide support, using an ODS-Hypersil column and a triethylammonium acetate buffer system, is shown in *Figure 3*. The oligodeoxyribonucleotide mixture in *Figure 3* had undergone a two-step deblocking procedure to remove the phosphate and nucleobase protecting groups.

Dissolve the residue in water and extract three times with an equal volume of ether as described in Section 4.1. Evaporate the remaining aqueous solution to dryness and dissolve in 0.5 ml of distilled water prior to injection on the 9.4 x 250 mm column of ODS-Hypersil. In contrast to either controlled pore glass or silica gel supports, oligodeoxyribonucleotides prepared on the Kieselgur-polydimethylacrylamide supports contain a relatively large amount of u.v.-absorbing material which cannot be entirely removed by extraction with ether and this material elutes near the void volume of the column. This does not, however, disturb isolation of the desired oligodeoxyribonucleotide. The early eluting peak of *Figure 3* also contains the failure sequences which lack the lipophilic 5'-terminal protecting group. In this case the gradient was started with 20% acetonitrile present in the mobile phase. Under these conditions the shorter failure sequences are eluted near the column's void volume.

It is generally desirable to monitor carefully the isolation procedure to determine whether a portion of the desired oligomer is lost as a result of the chromatographic isolation. This can be accomplished by analysis of a small aliquot of the partially deblocked oligonucleotide mixture on a 4.6 x 250 mm column using the conditions of *Figure 3* with a flow of 1.5 ml/min. Integration of this analytical

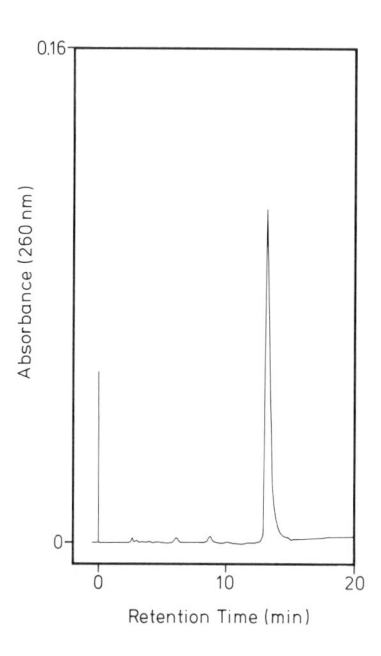

Figure 4. Analysis of the purified product d(CpGpTpTpTpTpGpC) by anion-exchange chromatography using a 4.6 x 250 mm column of APS-Hypersil (5 μm). Other conditions are as described in the legend to *Figure 2*.

chromatogram indicates that the peak eluting at 13 min contains 7.7% of the total eluted u.v.-absorbing material as monitored at 260 nm. In the total 0.5 ml sample 1215 A_{260} units were measured. With a 100% yield in the chromatographic step, 94 A_{260} units of oligodeoxynucleotide containing the 5'-terminal 9-phenylxanthenyl group should be expected. From the preparative isolation (*Figure 3*), 85 A_{260} units were obtained after removal of the volatile buffer. This represents a column yield of 89%.

To complete the isolation, dissolve the partially protected oligodeoxyribonucleotide in 2 ml of 80% acetic acid/water and keep at 0°C for 30 min. Evaporate the solution to dryness and co-evaporate the residue evaporated from distilled water three times. Dissolve the residue in 2 – 3 ml of distilled water and extract three times with an equivalent amount of diethyl ether. Evaporate the aqueous phase to dryness, dissolve the residue in 0.5 – 1.0 ml distilled water and lyophilise. In this case the procedure resulted in 70 A_{260} units of (dCpGpTpTpTpTpGpC).

The above chromatographic separation occurs primarily as a function of hydrophobic interactions. In order to confirm that a single species is present, a small aliquot of the final solution should be re-analysed on an anion-exchange column (APS) as illustrated in *Figure 4*.

The isolated yield of 70 A_{260} units is roughly 1.0 μmol (ϵ_{260} = 67 000). The synthesis began with 3.3 μmol of polymer-bound deoxyribonucleoside based upon the spectroscopic measurement of the 9-phenylxanthenyl cation at 375 nm (32 A_{375}/μmol). The penultimate coupling indicated that the synthesis produced

Chromatographic Purification of Synthetic Oligonucleotides

approximately 2.6 µmol of fully protected oligodeoxyribonucleotide. This represents a 79% assembly efficiency or an average 96% coupling efficiency per step. However, after isolation of the final product, a 32% overall yield is indicated. We have tried on numerous occasions to monitor the isolation of a particular oligodeoxyribonucleotide product and while some material is lost as a result of the chromatographic procedures it does not account for the difference observed between final yields of fully deblocked product and assembly efficiencies based on spectroscopic measurements of the 4,4'-dimethoxytrityl or 9-phenylxanthenyl cation.

Finally, it is possible to isolate chemically synthesised oligodeoxynucleotides by methods other than the chromatographic procedures described in this chapter. For small quantities of material, polyacrylamide gel electrophoresis can be successfully used, as described in Chapters 3 and 6. For larger quantities of material, there are at least two reports (4,5) indicating that t.l.c. using silica gel plates can be used successfully to isolate oligodeoxyribonucleotides in surprisingly high yields. It is also possible with relatively short fragments (decamers or shorter) to use gel-permeation chromatography on polysaccharide or polyacrylamide gel columns.

4.3 Analysis of Oligodeoxyribonucleotide Products

To obtain both sequence information and indication of purity, the wandering spot analysis as described in Chapter 6 is probably most effective. It requires, however, a reasonable amount of time and the use of radioisotopes. In order to obtain information concerning purity relatively quickly, a simple deoxyribonucleoside analysis can be employed. Digest a small aliquot of oligodeoxyribonucleotide completely using the following mixture at 37°C for a minimum of 4 h.

0.5 A_{260} unit oligodeoxyribonucleotide
50 mM Tris·HCl pH 8.0
10 mM $MgCl_2$
3 units snake-venom phosphodiesterase (Boehringer)
3 units bacterial alkaline phosphatase (Sigma).

Analyse an aliquot of this mixture by reversed phase chromatography using the following conditions.

Column: 4.6 x 250 mm ODS-Hypersil (5 µm)
Buffer A: 20 mM KH_2PO_4 (pH 5.5)
Buffer B: 20 mM KH_2PO_4 (pH 5.5), 10% CH_3OH
Gradient: 0−50% Buffer B in 30 min
Flow: 1.5 ml/min
Temperature: 35°C.

The retention times will depend on the age of the column. In a typical example, retention times of 4.3 min (dC), 10.1 min (dG), 13.0 min (dT) and 20.8 min (dA) were recorded. The peak areas resulting from the analysis can be integrated and the relative content of each deoxyribonucleoside determined. For the mobile phase conditions shown above extinction coefficients at 260 nm of 7700 (dC), 11 500 (dG), 8830 (dT) and 15 200 (dA) per mmol have been determined. An electronic integrator is the easiest way to obtain the desired areas. It is, however, also possible to increase the chart speed (to 10 cm/min) of a normal recorder and subsequently cut out and weigh the peaks resulting from the deoxyribonucleoside analysis.

In some cases the oligodeoxyribonucleotide can be digested only with snake-venom phosphodiesterase. This will produce 2'-deoxyribonucleoside-5'-phosphates and one 2'-deoxynucleoside resulting from the deoxynucleoside residue of the 5'-terminal position. It is then necessary to use a mobile phase containing an ion-pairing reagent in order to resolve the four 2'-deoxynucleoside-5'-monophosphates and four 2'-deoxynucleosides using reversed-phase chromatography (6). In our experience the resolution of thymidine-5'-monophosphate from 2'-deoxyguanosine-5'-monophosphate is difficult to reproduce in order to obtain an accurate integration of the peak areas. To avoid this problem we generally analyse oligodeoxyribonucleotide products using the simple 2'-deoxyribonucleoside analysis described above.

4.4 Isolation of Oligoribonucleotides

In contrast to DNA fragments, the chemical synthesis of RNA fragments is generally performed in solution (see Chapter 7). Intermediate fully protected oligoribonucleotides are purified using silica-gel chromatography after each coupling step. In this case a pure oligoribonucleotide species is present prior to the deprotection steps. This considerably simplifies the isolation procedure since now it is only necessary to separate the fully deblocked oligomer product from the reagents used in the deprotection steps as well as the phosphate, nucleobase and ribose protection groups. This can generally be easily accomplished using soft gel anion-exchange supports (Sephadex A-25) or gel-permeation supports (Sephadex G-50). These methods are well established and will not be discussed further here. There is, however, some value in using a chromatographic technique such as h.p.l.c. to monitor the isolation as well as to analyse product purity.

4.4.1 *Isolation of a Pentaribonucleotide*

In this example the fully protected pentamer CpUpGpUpG was prepared by chemical synthesis in solution using bifunctional phosphorylating reagent 2-chlorophenyl-O,O-bis(1-benzotriazolyl) phosphate (7).

(i) Since RNA fragments are sensitive to ribonucleases (such as those found on human skin), make up all aqueous solutions using de-ionised water which has been distilled twice from glass.

(ii) Heat all glass equipment which will be used in the isolation at 150°C for 18 h. Disposable polyethylene gloves can also be used during the deprotection procedure.

(iii) Deblock the fully protected pentamer, 100 mg (31 μmol) in three consecutive steps. First remove the phosphotriesters with *syn*-pyridinecarboxaldoxime in the presence of tetramethylguanidine, deprotect the nucleobases with concentrated ammonia and the ribose residues with hydrochloric acid pH 2.0 (see Chapter 7).

[Chromatogram figure]

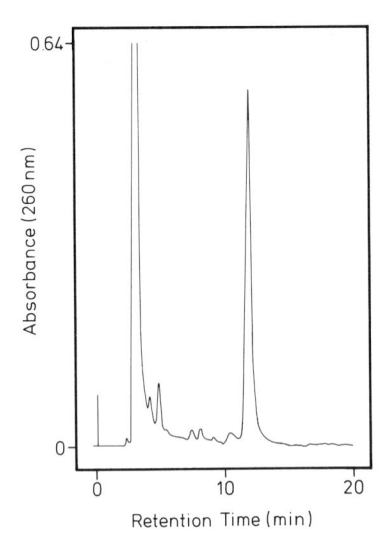

Figure 5. Analysis of the crude mixture containing the fully deblocked pentamer CpUpGpUpG on a 4.6 x 250 mm column of APS-Hypersil (5 μm): buffer A: 0.05 M KH_2PO_4 (pH 4.5); buffer B: 0.9 M KH_2PO_4 (pH 4.5), 10% methanol; gradient: 0–100% buffer B in 60 min; flow 1.5 ml/min, temperature 35°C.

(iv) Adjust the pH of the solution to 7.0 and analyse a small aliquot of the mixture by anion-exchange chromatography using an APS column as illustrated in *Figure 5*. The pyridinaldoxime and the hydrolysed blocking groups elute at the column void volume. The fully deblocked pentamer elutes with a retention time of 12 min. Integration of the analysis of *Figure 5* indicates that the pentamer comprises 19% of the total eluted material as monitored at 260 nm. The 60 ml crude mixture contains 5400 A_{260} units. With a 100% column yield 1026 A_{260} units would be expected.

(v) Chromatograph the mixture on a 2 x 25 cm column of Sephadex A-25 and develop with a gradient of 0.1–0.7 M triethylammonium bicarbonate (pH 7.5). In this case the product eluted at approximately 0.45 M salt.

(vi) In order to determine exactly which fractions from the Sephadex A-25 column should be pooled, analyse small aliquots (10–50 μl) from the leading and tailing portions of the peak of interest directly on the APS column using the conditions of *Figure 5*.

(vii) Pool fractions containing pure pentamer, evaporate to dryness, co-evaporate from distilled water a number of times and finally lyophilise.

The resulting 984 A_{260} units of the fully deblocked pentamer CpUpGpUpG indicates that only minimal material was lost during the chromatographic isolation in this case. The 21 μmol of fully deblocked pentamer represents a 68% yield for the three-step deprotection procedure.

4.5 Analysis of Oligoribonucleotides

Purity and sequence analysis is most effectively carried out using the wandering spot analysis. However, as with oligodeoxyribonucleotides, oligoribonucleotides can be rapidly analysed for nucleoside composition using h.p.l.c. In this case it is possible to resolve the four ribonucleosides and four ribonucleoside-3'-phosphates easily using a reversed phase column and a mobile phase containing a phosphate buffer and 1% methanol (8).

Digest oligoribonucleotides using the following mixtures at 37°C for a minimum of 2 h.

0.5 A_{260} unit oligoribonucleotide
10 mM sodium acetate pH 4.5
4 units ribonuclease T2 (Sankyo through Koch-Light Laboratories).

An aliquot from this mixture can then be analysed on a reversed phase column (ODS-Hypersil) using a mobile phase of 20 mM KH_2PO_4 pH 5.5 and 1% methanol. The order of elution using this mobile phase is: Cp, Up, C, Gp, U, Ap, G, A. In order that resolution of all eight species occurs, it is important that the pH of the mobile phase is carefully adjusted to pH 5.5. With old columns resolution between C and Gp is often lost, at which time the column should be repacked. Using the isocratic mobile phase described, a relatively simple h.p.l.c. system can be constructed for oligoribonucleotide analyses. On the other hand the elution of adenosine under these conditions requires more than 60 min. This can be considerably reduced if gradient elution is used. With this analysis the 3'-terminal residue will be eluted from the column as a nucleoside. The remaining residues will be 3'-monophosphates.

4.6 Isolation and Analysis of Oligoribonucleotides Produced by Enzymatic Syntheses

There are a number of polymerases and ligases which can be used to synthesise enzymatically both DNA and RNA fragments. This section will only discuss the analysis and isolation of mixtures produced by the action of T4 RNA ligase, the use of which is described in Chapter 8. The use of T4 RNA ligase results in the production of a phosphodiester linkage between an acceptor oligonucleotide containing a free 3'-hydroxyl and a donor oligonucleotide containing a 5'-terminal phosphate. During the course of the reaction the donor molecule is activated to the corresponding 5'-adenylylated species with ATP. Upon formation of the phosphodiester bond AMP is released. Since reactions involving the enzyme often do not go to completion the reaction mixture contains varying amounts of acceptor, donor, adenylylated donor, AMP, ATP, product and sometimes various side products. Polyacrylamide gel electrophoresis can be used to isolate small quantities of the desired oligonucleotide product. However, for larger quantities, chromatographic methods, primarily those involving h.p.l.c. are preferable. Using h.p.l.c., analysis it is also possible to monitor two consecutive enzymatic reactions such as formation of the 5'-phosphorylated donor using ATP and T4 polynucleotide kinase and its subsequent reaction with an acceptor in the presence of ATP and T4 RNA ligase. This will be exemplified for the analysis and subsequent isolation of the synthesis of the decamer GpCpGpGpApUpUpUpApm²Gp (m²G is N^2-methylguanosine).

Chromatographic Purification of Synthetic Oligonucleotides

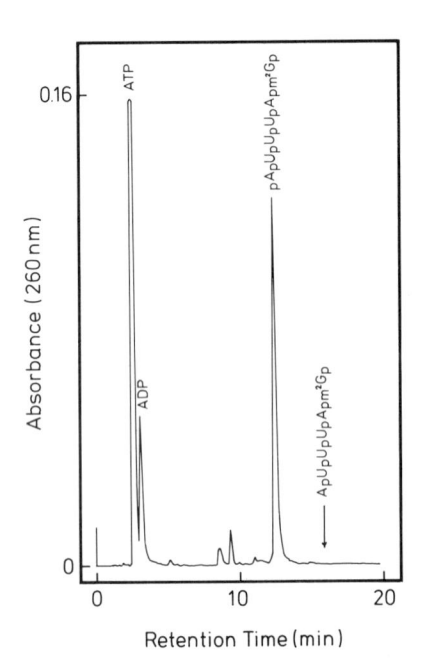

Figure 6. Analysis of the phosphorylation of the hexamer ApUpUpUpApm²Gp using ATP and T4 polynucleotide kinase on a 4.6 x 250 mm column of ODS-Hypersil (5 μm): buffer A: 0.02 M KH$_2$PO$_4$ (pH 5.5); buffer B: 0.02 M KH$_2$PO$_4$ (pH 5.5), 70% methanol; gradient: 0–100% buffer B in 60 min; flow 1.5 ml/min; temperature 35°C.

GpCpGpG was prepared chemically using a bifunctional phosphorylating reagent activated by 1-hydroxybenzotriazole (7) as described in Chapter 7. ApUpUpUpApm²Gp was isolated from a ribonuclease T1 hydrolysis of tRNA specific for the amino acid phenylalanine of yeast. As a result of the action of the ribonuclease, the hexamer was isolated with a 3'-terminal phosphate which will act as a blocking group for the 3'-hydroxyl moiety of the donor molecule. Two enzymic reactions are then required to prepare the desired product. Prepare the 5' phosphorylated oligomer pApUpUpUpApm²Gp from the isolated hexamer, ATP and T4 polynucleotide kinase. Analyse a small aliquot from this reaction mixture after a 2 h incubation at 37°C (*Figure 6*) using a reversed phase support (ODS-Hypersil) and a methanol gradient in phosphate buffer. One advantage of this system is that very polar molecules such as ATP and ADP elute very quickly from the column and there is little chance they will effect the resolution or identification of the oligonucleotide substrates and products. (In this case the analysis of *Figure 6* confirmed that the hexamer had been quantitatively phosphorylated). Then heat the reaction mixture at 95°C for 2 min to inactivate the polynucleotide kinase and add the acceptor (GpCpGpG) and T4 RNA ligase (see Chapter 8).

In this case after 48 h at 17°C a small aliquot of the reaction mixture was again analysed (*Figure 7*) using the conditions of *Figure 6*. Integration of the chromatogram of *Figure 7* indicates that the reaction was 70% complete. The product oligomer is well resolved from both starting materials and could therefore be isolated preparatively on a larger column using the phosphate-methanol mobile

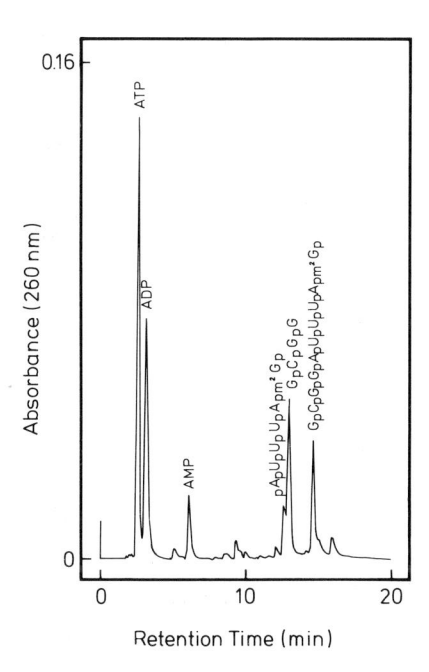

Figure 7. Analysis of the reaction between GpCpGpG and pApUpUpUpApm²Gp as catalysed by T4 RNA ligase on a 4.6 x 250 mm column of ODS-Hypersil (5 μm). Other conditions are as listed in the legend to *Figure 6*.

phase. However, a subsequent desalting step would be required. Isolate the product using a volatile buffer and the following conditions.

  Column: 9.4 x 250 mm ODS-Hypersil (5 μm)
  Buffer A: 0.1 M triethylammonium acetate (pH 7.0)
  Buffer B: 0.1 M triethylammonium acetate (pH 7.0), 80% acetonitrile
  Gradient: 0−100% BufferB in 60 min
  Temperature: 35°C
  Flow: 4 ml/min

Using this volatile mobile phase the solutes are eluted according to the number of phosphate groups present. The product (decamer) remains the last peak, is well resolved and can be collected in high purity. Evaporate the collected material to dryness, co-evaporate a number of times from water and finally lyophilise. In this case 35 A_{260} units (0.25 μmol) of product were isolated by this procedure. Nucleoside-nucleotide analysis was carried out as described in Section 4.5.

Mixtures produced by the action of T4 RNA ligase can also be analysed by chromatography on anion-exchange columns as described fully elsewhere (8).

5. MIXED MODE CHROMATOGRAPHY

Nucleic acids and oligonucleotides are essentially polyanionic chains containing lipophilic nucleobases. Both anion-exchange as well as hydrophobic interaction chromatography can be used for separation and isolation of a particular species.

A third possibility would be to use a mixed mode chromatographic support where both the ionic and hydrophobic interactions occur. A number of modified, bonded phase, silica matrices have been made with this aim in mind (9,10,11). In one approach an anion-exchange stationary phase (APS-Hypersil) has been covalently modified with various organic hydrophobic moieties to produce a series of hydrophobic anion-exchange supports. In a second approach a hydrophobic stationary phase (ODS-Hypersil) has been coated with a tetraalkylammonium salt to introduce sites for the desired ionic interactions. Using a mobile phase of high salt concentration the tetraalkylammonium salt is essentially irreversibly bound to the hydrophobic support. The former supports have been valuable in the resolution of sequence-isomeric oligonucleotides of the same chain length as well as some nucleic acid mixtures. The latter supports can be used to isolate tRNAs of high purity. In some cases partial hydrolysates of homopolymers can be resolved even up to long chain length [e.g., $(Up)_{89}$ from $(Up)_{90}$]. It is possible that such mixed mode chromatography will be useful for the isolation of very large synthetic DNA or RNA fragments but we have not yet prepared any oligonucleotides larger than 20 residues, which would be required to examine the resolution available with columns containing mixed mode matrices.

6. CONCLUSIONS

Chromatographic analysis and isolation of oligodeoxyribonucleotides or oligoribonucleotides prepared by chemical or enzymatic syntheses is generally best accomplished using h.p.l.c. Beyond the initial investment required for an h.p.l.c. system the most expensive items are the columns, which must be renewed at regular intervals. These costs can be drastically reduced by learning to self-pack columns as described in Appendix II. We recommend the preparation or purchase of four columns for a general approach to analysis and isolation: APS-Hypersil 4.6 x 250 mm, Whatman SAX 4.6 x 250 mm, ODS-Hypersil 4.6 x 250 mm and ODS-Hypersil 9.4 x 250 mm.

Chromatograms illustrated in this chapter have been produced on columns packed in the authors' laboratory (see Appendix II) using stainless steel fittings from Swagelok (Crawford Fitting Co., Cleveland, Ohio, USA). Similar prepacked columns or chromatographic supports in bulk are available from Whatman Inc., Clifton, New Jersey, USA or Maidstone, Kent, UK (Whatman SAX) and Shandon Southern Inc., Sewickley, Pennsylvania, USA or Runcorn, Cheshire, UK (APS-Hypersil and ODS-Hypersil).

7. REFERENCES

1. Marugg,J.E., McLaughlin,L.W., Piel,N., Tromp,M., van der Marel,G.A. and van Boom,J.H. (1983) *Tetrahedron Lett.*, **24**, 3989.
2. Gait,M.J., Matthes,H.W.D., Singh,M., Sproat,B.S. and Titmas,R.C. (1982) *Nucleic Acids Res.*, **10**, 6243.
3. Fritz,H.-J., Belagaje,R., Brown,E.L., Fritz,R.H., Jones,R.A., Lees,R.F. and Khorana,H.G. (1978) *Biochemistry (Wash.)*, **17**, 1257.
4. Chou,S.H., Hare,D.R., Wemmer,D.E. and Reid,B.R. (1983) *Biochemistry (Wash.)*, **22**, 3037.

5. Hare,D.R., Wemmer,D.E., Chou,S.H., Drobny,G. and Reid,B.R. (1984) *J. Mol. Biol.,* **171,** 319.
6. Fritz,H.-J., Eick,D. and Werr,W. (1982) in *Chemical and Enzymatic Synthesis of Gene Fragments*, Gassen,H.G. and Lang,A. (eds.), Verlag Chemie, Weinheim, p. 199.
7. Wreesman,C.T.J., Fidder,A., van der Marel,G.A. and van Boom,J.A. (1983) *Nucleic Acids Res.,* **11,** 8389.
8. McLaughlin,L.W. and Romaniuk,E. (1982) *Anal. Biochem.,* **124,** 37.
9. Bischoff,R., Graeser,E. and McLaughlin,L.W. (1983) *J. Chromatogr.,* **257,** 305.
10. Bischoff,R. and McLaughlin,L.W. (1983) *J. Chromatogr.,* **270,** 117.
11. Bischoff,R. and McLaughlin,L.W. (1984) *J. Chromatogr.,* in press.

CHAPTER 6

Purification and Sequence Analysis of Synthetic Oligodeoxyribonucleotides

RAY WU, NAI-HU WU, ZAHRE HANNA, FAWZY GEORGES AND SARAN NARANG

1. INTRODUCTION

The progress made in recent years in the chemical synthesis of oligodeoxyribonucleotides has contributed significantly to the present state of the art. The two most effective methods are phosphotriester (1) (Chapter 4) and phosphite-triester (phosphoramidite) (2) (Chapter 3). Both methods can deliver high coupling efficiencies (~95%) and allow synthesis of oligodeoxyribonucleotides at fast speeds. Synthetic oligodeoxyribonucleotides of well-defined sequences have a number of important applications in molecular biology and recombinant DNA research (3); they are outlined in Chapter 1.

Within the last year, several types of DNA synthesis machine have become available, which are capable of assembling oligonucleotides of 15 – 50 residues in length in less than 24 h using solid-phase methods. Although there are occasional failures due to mechanical or chemical problems, these machines will have a great impact on molecular biology, because they will make oligonucleotides available to a large number of investigators. If large amounts of oligonucleotides (>50 mg) are required for biophysical studies, such as X-ray crystallography or n.m.r., they are best prepared by applying solution chemistry using the phosphotriester method (1), since the capacity of the machines is still rather low and scale-up of the solid-phase method has not been fully explored. However, for most purposes, the amounts that can be prepared using the machines are adequate.

Once the oligodeoxyribonucleotides have been prepared, it is essential that their sequences are confirmed by direct sequence analysis. There are two rapid methods for determining the sequence of an oligodeoxyribonucleotide in the 10 – 25 nucleotide size range.

Method 1. This method is based on partial enzymatic digestion of a single end-labelled oligodeoxyribonucleotide to generate a family of oligonucleotides of varying length, which is then fractionated by two-dimensional electrophoresis-homochromatography. The principle of this mobility-shift or wandering-spot method was first developed for oligoribonucleotides to distinguish the mobility-shift among three different nucleotides (Up, Cp, Ap) (4) and was then applied to oligodeoxyribonucleotides to distinguish among two (5), three (6) or four (7) nucleotides. Inspection of the mobility-shift of adjacent nucleotides in a sequence generally allows an experienced worker to deduce the sequence with reasonable reliability. The application of a formula for the calculation of the mobility of oligodeoxyribonucleotides can increase the reliability (8). If an oligodeoxyribonucleotide is sequenced from both ends by using snake venom phosphodiesterase digestion of a 5′-labelled oligomer, and then separately using spleen phosphodiesterase digestion of the same oligomer labelled at the 3′ end, total reliability of the sequence can be assured (7,9). The sequence analysis of several oligomers can be carried out in parallel and completed in one day.

Method 2. This method is based on chemical modification of an oligodeoxyribonucleotide, followed by cleavage to produce a family of products of varying length, which is fractionated by polyacrylamide gel electrophoresis (10). This method is suitable for a long DNA molecule, but often misses a few nucleotides close to the labelled end. A modification of this method (11) appreciably improved the sequence analysis of short oligodeoxyribonucleotides. However, there are often problems in obtaining reliable results, and we found that certain steps are not always reproducible. We have adopted many of the details for the cleavage reactions from this method and improved it in several places. Ambiguities occur in sequence determination when one uses only four standard cleavage reactions (G, A+G, C, C+T) and even five reactions (including A>C) (10). Banaszuk *et al.* (11) included a potassium permanganate reaction for T-specific cleavage (12). However, we have found that the osmium tetraoxide modification of thymine (13) is more reproducible than the permanganate reaction, although in our experience it is necessary to cut down the reaction time of osmium tetraoxide treatment from 15 min to 3 min. Also included in this chapter is an earlier method for the labelling and purification of oligodeoxyribonucleotides (14). The combined labelling and sequencing method presented here is easy to carry out and gives reproducible results.

In most cases, the sequence of a 5′-labelled oligodeoxyribonucleotide of less than 15 residues in length can be unambiguously determined by either of the two above methods. However, in certain cases there are ambiguities and the sequence needs additional verification. One can determine the sequence of the same oligonucleotide labelled at the 3′ end by the enzymatic method (method 1) or by the chemical method (method 2), or alternatively one can determine the sequence of a 5′-labelled oligonucleotide using both the chemical cleavage method and the enzymatic cleavage method. The first method is especially suitable for determining the first six nucleotides from the 5′-labelled end, and the second is especially suitable for analysing nucleotides 4−40 residues in length. The choice of method also depends on the nature of the ambiguity. For example, if the chemical method is used and the ambiguity is in deciding whether a particular nucleotide is C or T, then method 1 should be used, since the enzymatic method is especially suitable for resolving C/T ambiguity. In general, method 1 is preferred for oligonucleotides shorter than 15 residues, and method 2 is preferred for oligonucleotides longer than 20 residues.

In this chapter, we emphasise the technical aspects of sequencing oligodeoxyribonucleotides in order that other investigators can readily follow the protocols and carry out analyses. A more detailed discussion of DNA sequencing in general can be found in an excellent book by Hindley (15).

2. PURIFICATION PROCEDURES

After the removal of all the protecting groups, the desired oligomer has to be purified through one or two efficient chromatographic procedures. These include:
(i) t.l.c. on polyethyleneimine (PEI) cellulose (16);
(ii) h.p.l.c. (see Chapter 5);
(iii) t.l.c. on silica gel plates;
(iv) preparative polyacrylamide slab gel electrophoresis.

Both methods (i) and (ii) can fractionate oligonucleotides of up to 17 residues. Method (iii) can fractionate longer oligonucleotides but the resolutions are not as good as (i) or (ii). Method (iv) can also fractionate longer oligonucleotides but with much better resolution. In our experience the sequential use of t.l.c. on silica-gel plates, followed by polyacrylamide gel electrophoresis is the most convenient and economical combination.

2.1 T.l.c. on Silica Gel Plates

Oligodeoxyribonucleotides can be purified on silica gel 60 F_{254} t.l.c. plates (17) as follows (see also Appendix I).

Analytical pre-coated silica gel 60 F_{250} plates (0.25 mm) on glass (20 x 20 cm) are supplied by Brinkman or by E.Merck.

(i) Apply the crude reaction mixture containing up to 200 μg of total oligonucleotide mixture in 100 μl of aqueous pyridine on one plate as a uniform 15 cm band at the origin.
(ii) Carefully blow off excess solvent using a cool hair dryer and develop the plate in n-propanol/concentrated ammonia/water (55:35:10 v/v). Elution time is 8 – 15 h.
(iii) After once again blowing off solvent observe the plate under u.v. light **(WARNING: eye protection required)**. Generally the slowest migrating major band is the desired oligonucleotide.
(iv) Scrape off the silica gel corresponding to this band into an Eppendorf centrifuge tube and consecutively elute the oligonucleotide from the silica with a mixture of ethanol and water (3:1 v/v) in 300 μl, 100 μl and 100 μl volumes, respectively, by soaking for 1 h each time, followed by centrifugation and collection of the supernatant.
(v) Remove the solvent under vacuum with occasional shaking. The product should be a white foamy material.

2.1.1 *Advantages*
(i) This system is very economical and easy to work with as compared with h.p.l.c., which requires expensive equipment and considerable skill of operation.
(ii) Large-scale purification is possible.
(iii) Recovery of the desired compound is simple.
(iv) Fragments of oligodeoxyribonucleotides up to 30 residues have been purified using this system.

2.1.2 Disadvantages

(i) There is no well-defined correlation of the mobility of variously sized fragments, and sometimes it is difficult to locate the desired band.
(ii) It is possible to isolate a wrong band if the desired one is not the major band.
(iii) Confusion is most likely to occur when there has been an inefficient synthesis due to errors or the use of impure reagents in oligonucleotide assembly.

It is therefore always important to characterise the oligonucleotide by sequencing, after first labelling it with a radioactive phosphate group.

2.2 Polyacrylamide Slab Gel Electrophoresis

In some cases where the silica gel system has failed to resolve the reaction mixture adequately, or if ultra-pure oligonucleotides are needed, a slab polyacrylamide-7 M urea gel should be used in addition to the silica gel system. The identification of the desired compound on a polyacrylamide gel is usually unambiguous, due to the fractionation of oligonucleotides based on size. There are two common observations worth pointing out.

(1) Mobility of the 5' hydroxyl-containing oligomer is always slower than its 5' phosphate derivative.
(2) Mobility of some fragments is dependent on the composition of the base sequence rather than its length. For example, the dodecamer sequence, d(A-A-A-C-G-C-G-A-A-A-G-C) moves much faster than several other dodecamers synthesised in our laboratory. This abnormal mobility may be due to stacking or formation of secondary structure of the GC-rich regions.

The method is as follows:

(i) Purify labelled oligodeoxyribonucleotides on 20% polyacrylamide gels containing 7 M urea (14) (cf. Chapter 3 where polyacrylamide gel purification of unlabelled oligonucleotides is described).
(ii) After electrophoresis at 800 V for 1 h, the radioactive oligonucleotides should be located by autoradiography (usually a 10 min exposure time is sufficient).
(iii) Cut the gel slice (usually the slowest moving band) carrying the radioactive oligonucleotide out of the gel with a razor blade.
(iv) Place the gel slice in a 5 ml disposable syringe (without a needle) and crush into fine pieces by pushing the gel slice through the syringe.
(v) Collect the gel pieces in an Eppendorf tube and soak in 2 – 3 volumes of extraction buffer [300 mM NaCl, 10 mM Tris-HCl (pH 8), 1 mM EDTA] for 12 h.
(vi) Centrifuge the tube at 4°C for 10 min in an Eppendorf centrifuge and pipette the supernatant into another Eppendorf tube.
(vii) Add two volumes of extraction buffer to the gel slices and incubate at room temperature for 3 h.

(viii) Centrifuge the tube and collect the supernatant.
(ix) Combine the supernatants and extract twice with one volume of phenol/chloroform/isoamyl alcohol (24:24:1 by vol.) containing 0.1% 8-hydroxyquinoline.
(x) Extract the aqueous layer three times with equal volumes of ether and finally precipitate with 2.5 – 3 volumes of 95% ethanol at −20°C for 4 h (recovery >70%).
(xi) Pellet the oligonucleotide by centrifugation at 4°C for 15 min in an Eppendorf centrifuge.

3. METHODS FOR LABELLING OLIGODEOXYRIBONUCLEOTIDES

3.1 Labelling of 5' Ends of Oligodeoxyribonucleotides with T4 Polynucleotide Kinase and [γ-^{32}P]ATP

For 5' labelling of an oligodeoxyribonucleotide (14) prepare the following mixture in a 1.5 ml Eppendorf tube:

4 μl [γ-^{32}P]ATP (specific activity ~2000 Curies per nmol obtained from Amersham; 20 pmol, 40 μCi) evaporated to dryness in a Buchler Vortex Evaporator (~15 min); 5 μl 2 x kinase buffer (120 mM Tris-HCl, pH 8.5, 20 mM MgCl$_2$, 20 mM dithiothreitol); 2 μl unlabelled 10 μM ATP (recently diluted from a 10 mM stock, pH 7.5), 20 pmol; 2 μl synthetic oligodeoxyribonucleotide, 20 pmol; 1 μl T4 polynucleotide kinase, 10 units, (obtained from B.R.L.).

Incubate at 37°C for 30 min. Add 2 μl hexokinase buffer (50 mM Tris-HCl, pH 7.5, 5 mM ATP, 5 mM MgCl$_2$, 10 mM glucose, 50 μg/ml of hexokinase (Boehringer).

Incubate at 37°C for 10 min. (N.B. This step can often be omitted.)

Purify the [^{32}P]oligonucleotide away from [γ-^{32}P]ATP by DE-81 paper chromatography as follows:

(i) Cut a 1 x 30 cm strip of DE-81 paper (Whatman), and spot 5 μl of 10 mM ATP, 100 mM EDTA (pH 7) along a line (origin) 6 cm from one end of the strip. Allow it to dry at room temperature.
(ii) Spot the reaction mixture from the labelling experiment along the same origin line, 5 μl at a time and allow it to dry.
(iii) Place the DE-81 paper strip inside the solvent trough of a paper chromatography jar with a glass rod lifting the paper above the edge of the trough and a second glass rod weighing down the paper.
(iv) Add 50 ml of 0.2 M ammonium formate solution to the trough, cover the chromatography jar, and allow the chromatogram to develop for 1 h. The solvent front is usually about 5 cm from the lower edge of the paper.
(v) Lift the paper up by the lower edge and hang it upside down to dry.
(vi) Cut the paper into seven pieces (1.2 cm for the first piece which includes the origin, 1.2 cm for the next piece, 3 cm each for the next five pieces).
(vii) Place each piece of paper inside a scintillation vial, and count for 0.1 min for each sample using Cerenkov counting. If the polynucleotide kinase reaction has proceeded well, over 30% of the counts should be at the origin

(60% conversion), which includes the [^{32}P]oligodeoxyribonucleotide. The yield of the labelled oligonucleotide should be around 5 x 10^6 c.p.m.

(viii) The labelled oligonucleotide should be eluted from the paper strip in a 0.5 ml Eppendorf tube, which has been pierced at the bottom with a needle.

(ix) Roll up the DE-81 paper carrying the labelled oligodeoxyribonucleotide and place it inside the tube.

(x) Add 10 μl of 1 M triethylammonium bicarbonate solution (TEAB) (pH 9.5) and close the cap of the 0.5 ml tube.

(xi) Place the tube inside a 1.5 ml Eppendorf tube and centrifuge the tubes in an Eppendorf centrifuge at 4°C for 1 min.

(xii) Add 10 μl of 1 M TEAB to the 0.5 ml tube and repeat the elution twice.

(xiii) Evaporate the eluate from the three elutions to dryness (in the original 1.5 ml tube) using a Buchler Vortex Evaporator. This takes about 30 min.

(xiv) Dissolve the radiolabelled oligonucleotide in 15 μl of glass-distilled water and evaporate to dryness. Repeat this step three times.

(xv) Dissolve the labelled oligonucleotide (over 80% recovery) in 100 μl of TE buffer [5 mM Tris-HCl (pH 8), 0.5 mM EDTA] containing 80 μg of sonicated calf thymus DNA (prepared by sonication of a 1 mg/ml solution at 0°C, five times for 30 sec each, extracted three times with re-distilled phenol and twice with ether, and dialysed against 20 volumes of TE buffer).

(xvi) Count 2 μl of the solution (it should contain ~1 x 10^5 c.p.m.).

(xvii) Keep the solution of labelled oligonucleotide in two tubes at −20°C; it is stable for over a week. Use approximately 100 000−150 000 c.p.m. for each chemical cleavage reaction, and 500 000 c.p.m. for the enzymatic cleavage reaction.

3.2 Labelling of 3' Ends of Oligodeoxyribonucleotides with Terminal Transferase and [α-^{32}P]Cordycepin-5'-Triphosphate

For 3' labelling of an oligodeoxyribonucleotide (9) prepare the following mixture in a 1.5 ml Eppendorf tube:

10 μl [α-^{32}P]cordycepin-5-triphosphate (~5000 Curies per nmol obtained from New England Nuclear; 20 pmol, 100 μCi) in 5 mM Tricine (pH 7.6), evaporated to dryness;

10 μl 1.4 x transferase buffer (200 mM potassium cacodylate, pH 7.0, 0.2 mM dithiothreitol, 1.5 mM CoCl$_2$);

2 μl synthetic oligodeoxyribonucleotide, 20 pmol;

2 μl terminal transferase (P.L. Pharmacia), 2 units.

Incubate at 37°C for 30 min. Add 10 μl 0.5 M ammonium acetate, and 80 μl 95% ethanol. Mix, chill at −70°C for 20 min, centrifuge for 15 min in an Eppendorf centrifuge and remove supernatant. Purify the [^{32}P]oligonucleotide away from [α-^{32}P]cordycepin-5'-triphosphate by adding 10 μl of water to the pellet, mixing and spotting the sample on a 1 x 30 cm strip of DE-81 paper and continue as in Section 3.1.

4. SEQUENCE DETERMINATION BY ENZYMATIC DEGRADATION AND TWO-DIMENSIONAL CHROMATOGRAPHY

A method for sequence analysis of oligodeoxyribonucleotides based on the characteristic mobility shift of their sequential partial enzymatic degradation products by a two-dimensional chromatography procedure has been described (4,6,7,18). For oligonucleotides between 1 and 9 residues in length, visual inspection of the mobility-shift is usually adequate to deduce the sequence. Oligonucleotides between 10 and 25 residues in length may need the application of a formula for calculating the electrophoretic mobilities, and then comparison of these with the observed mobility-shift values to deduce the sequence accurately (8). It is most convenient to label the 5' end of an oligonucleotide with ^{32}P and use snake venom phosphodiesterase to partially digest it to give a family of sequential degradation products. The sequence analysis is most reliable near the 5'-labelled end and becomes less and less reliable toward the 3' end, especially in regions where the sequence is rich in A and G and the size of the oligonucleotide is longer than 10 or 15 residues. When ambiguity is encountered it may be necessary to label the oligonucleotide at the 3' end with either a [^{32}P]ribonucleotide (19), or more conveniently with [^{32}P]cordycepin phosphate (9) using deoxynucleotidyl terminal transferase. The 3'-labelled oligonucleotide is then partially digested with spleen phosphodiesterase and the family of degradation products fractionated by two-dimensional chromatography (7) to determine reliably the sequence near the 3' end.

The protocol for determining the sequence of a 5' [^{32}P]oligonucleotide by the mobility-shift (or wandering-spot) method is as follows.

Mix together:
10 μl 5'-[^{32}P]oligonucleotide (5 x 10^5 c.p.m.);
10 μl 40 mM Tris-HCl (pH 8.3) and 20 mM MgCl$_2$;
4 μl carrier tRNA (5 mg/ml) (Becton Dickinson, *E. coli* tRNA).

Chill tube to 0°C for 5 min and add 2 μl snake venom phosphodiesterase (Worthington or Boehringer) (1 mg/ml). Incubate at 37°C and take aliquots (4 μl) at 0, 5, 10, 20 and 30 min. Using a micropipette, transfer each aliquot directly into another tube containing 20 μl of stop solution (1 N NH$_4$OH + 0.1 mM EDTA). Dry the combined digest *in vacuo* and then dissolve in 5 μl of water.

Fractionate the digest in the first dimension in 10% acetic acid containing 2 mM EDTA, adjusted to pH 3.5 with pyridine, on strips of cellulose acetate by high voltage ionophoresis using either flat-bed or up-and-over tanks. Pre-cut strips of cellulose acetate are available from Schleicher and Schüll in 3 x 55 cm long strips.

(i) Wet a cellulose acetate strip with fresh pH 3.5 buffer. To avoid the inclusion of air bubbles, this should be done from one side by flotation.

(ii) Place one end of the strip on the surface of the buffer contained in a Petri dish.

(iii) As it absorbs the buffer, wet the rest of the strip by passing it slowly across the surface of the liquid.

Purification and Sequence Analysis

(iv) Finally, completely immerse the strip in buffer to ensure it is completely wet. This buffer is also used in the buffer compartments of the electrophoresis tank.

(v) Blot the point of application, about 10 cm from one end of the strip, free of excess liquid and apply the digest as a small spot and allow it to soak in.

(vi) Apply a dye mixture of Xylene Cyanol FF (blue), acid fuchsin (red) and methyl orange (yellow) on each side of the digest. Care has to be taken to avoid the strip drying out while the sample is applied. Have the ends covered with wet tissue paper, and then remove.

(vii) Remove excess buffer from the strip by blotting and rapidly dipping in 'white spirit' (Varsol) to prevent evaporation of the buffer while the strip is put into the ionophoresis tank. The origin is at one end near the negative electrode vessel. Carry out ionophoresis at between 50 and 100 V/cm (usually 25 – 35 min at 3 – 5 kV).

(viii) Prepare DEAE-cellulose plates by taping plastic sheets of DEAE (Polygram Cel 300 DEAE/HR-2/15, Macherey Nagel, 20 x 40 cm) onto glass plates of the same size using double-sided tape.

(ix) Make pencil marks carefully on the edges of the plate 5 cm above the bottom, marking the place where the transfer is to be made. Blot nucleotides from the cellulose acetate onto the plates as follows.

(a) Remove the strip from the tank used for the first-dimensional run and monitor with a portable Geiger counter to find the position of the oligonucleotides (usually in the region between the red and blue markers).

(b) Place the strip on top of the DEAE-cellulose plate.

(c) Place three moist, but not too wet, strips of Whatman 3MM paper (2.5 x 25 cm) carefully over the cellulose acetate strip.

(d) Place a glass plate (5 x 20 cm) and a weight of about 0.5 kg on top so that the plate presses down evenly onto the cellulose acetate strip. Water thus flows from the 3MM paper, transferring the oligonucleotides onto the DEAE-celulose layer. It is preferable to use strips that still have some 'white spirit' on their surface, rather than let the buffer in the strip dry out, and to allow 20 – 30 min for the transfer. In order to ensure efficient transfer, the paper strips are wetted with water occasionally. When the dye markers are transferred onto the plate the transfer is usually complete.

(x) Take the plate and place the bottom end in a trough of water (2 cm deep).

(xi) Wait until the water front rises by capillary action to a level of about 1 inch above the location of the labelled oligonucleotides.

(xii) Clip a 3MM wick to the top of the plate using Terry clips (this is not always necessary and depends how far the plate is developed).

Carry out fractionation in the second dimension at 60°C in an oven using homomixture IV for 15 – 25 long or homomixture VI for 10 – 14 long (18) (partial hydrolysates of RNA). A tall tank is formed by inverting a thin-layer tank (20 x 20 cm) over another one (usually with slots to accept several plates). Make sure during chromatography that the joint between the two tanks is well greased to prevent evaporation of the solvent. Equilibrate the tank with about 100 ml of the homomixture at 60°C in the incubator for 1 h.

Place the wet plate immediately into the equilibrated tank and carry out ascending homochromatography until the front reaches the top of the plate (2–5 h).

Remove the plate, dry and apply radioactive ink in convenient locations to orient the plate. Autoradiograph in a cassette at −20°C or −70°C overnight.

The results of analysis of four oligonucleotides (I, II, III and IV) are shown in *Figure 1*. In reading the sequence, start from the upper left-hand corner. The first nucleotide is identified by comparing its electrophoretic mobility relative to that of the blue and the yellow dye markers. The sequence of the remaining nucleotides are deduced one at a time by comparing the mobility of the second nucleotide relative to the first, and so on, by applying the following general rules.

(i)   The addition of a dpC to any nucleotide or oligonucleotide always causes a relatively large electrophoretic mobility (first dimension) decrease.
(ii) The addition of a dpT always causes a relatively large electrophoretic mobility increase.
(iii) The addition of a dpA causes either no change or a small mobility decrease.
(iv) The addition of a dpG causes either no change or a small mobility increase.
(v)   The addition of a dpG or dpA causes a larger mobility decrease than that caused by dpC or dpT in the homochromatography dimension (second dimension).
(vi) The addition of a nucleotide to a mononucleotide causes a relatively larger shift than the addition of the same nucleotide to a di- or oligonucleotide.

For example, in deducing the sequence of oligonucleotide I (dodecamer) in *Figure 1*, the first radioactive spot represents a [^{32}P]dpT since the mobility is in between the blue and yellow markers and about two-thirds of the way toward the yellow marker. The second nucleotide is a dpT since the addition of this nucleotide causes a large electrophoretic mobility increase. The third nucleotide is also a dpT for the same reason. The fourth nucleotide is a dpG since it causes no change in the mobility (dpG usually causes no change in the mobility when the mobility of the oligonucleotide is already high before the addition of dpG, and it causes an increase in mobility when that of the oligonucleotide is relatively low prior to the addition of dpG). By following the above general rule, the sequence of the entire oligonucleotide is deduced one nucleotide at a time.

5. SEQUENCE DETERMINATION BY CHEMICAL CLEAVAGE REACTIONS AND GEL ELECTROPHORESIS

5.1 Base-specific Chemical Cleavage Reactions

Carry out reactions in 1.5 ml Eppendorf conical polypropylene tubes with attached snap-caps. (**WARNING: refer to the paper by Maxam and Gilbert (10) for the safe handling and storage of reagents for the cleavage reaction.**) In essence, dimethyl sulphate, hydrazine and piperidine should be stored in two lots: most of it in the original bottle, tightly capped, and 1 ml as a working solution in a small, screw-capped glass tube. The dimethyl sulphate working solution should be

Purification and Sequence Analysis

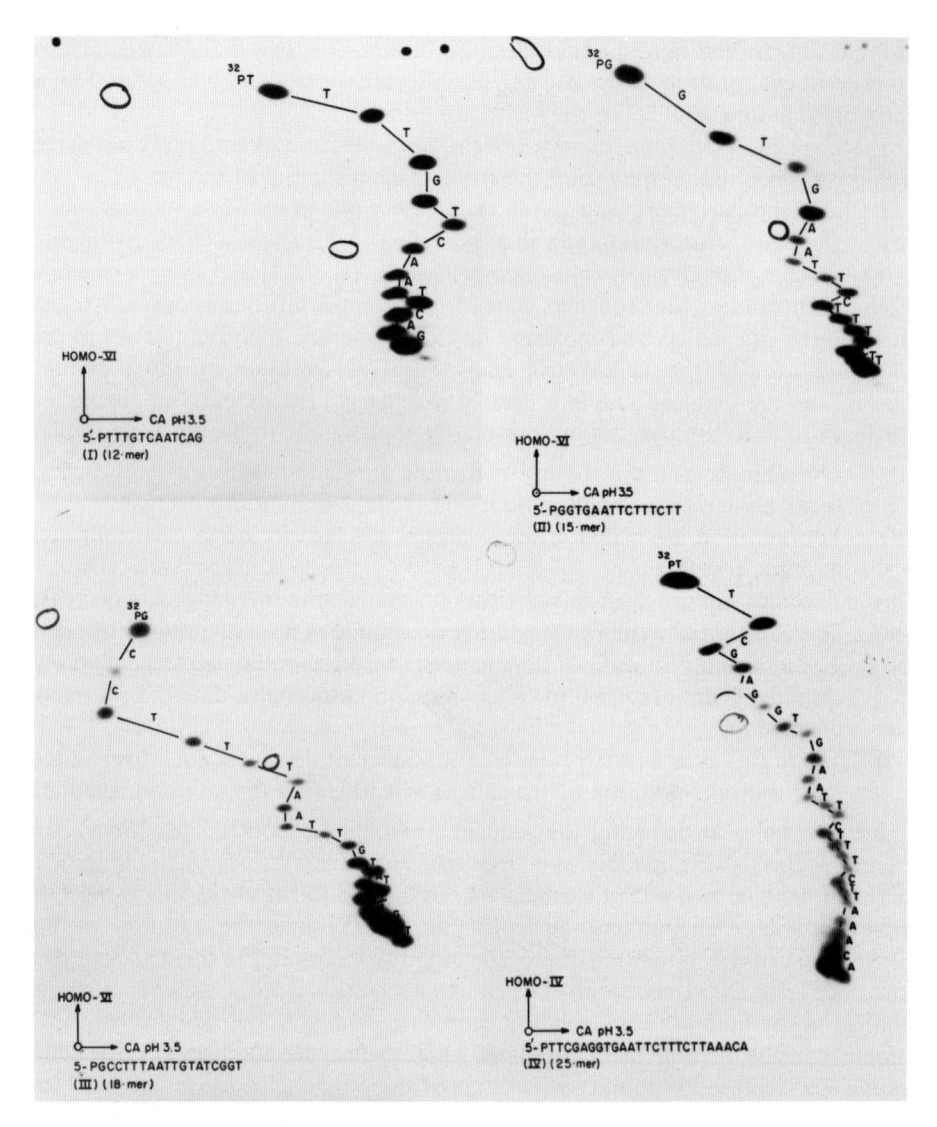

Figure 1. Sequence analysis of four synthetic oligodeoxyribonucleotides by the mobility-shift method.

replaced after several uses. The hydrazine working solution should be replaced every day.

In carrying out the limited cleavage reactions (first step of each reaction), one usually uses an incubation time which results in cleavage of approximately 50% of the starting oligonucleotide. If the cleavage reaction results in too few counts in the faster moving bands, the time for cleavage should be increased. If too few counts are found in the slower moving bands, the time for cleavage should be decreased.

5.1.1 *Limited Oligonucleotide Cleavage at Guanines (G)*

(i) Mix 200 µl of 50 mM sodium cacodylate (pH 8), 1 mM EDTA, 10 mM MgCl$_2$ with 5 µl of [^{32}P]oligonucleotide containing 1.5 µg calf DNA.

(ii) Chill on ice, and add 1 µl of dimethyl sulphate (99%), reagent grade.

(iii) Close the cap of the Eppendorf tube and mix (Vortex). Incubate at 37°C for 15 min. Chill to 0°C.

(iv) Add 50 µl of DMS stop solution [1.5 M sodium acetate (pH 7), 1 M mercaptoethanol], 5 µl of *E. coli* tRNA (10 mg/ml) and 750 µl of 95% ethanol (0°C).

(v) Close the cap and mix well (invert the tube four times). Chill at −70°C for 25 min in a dry ice-isopropanol bath. Warm up the sample until fluid and pellet the oligonucleotides in an Eppendorf centrifuge at 12 000 g for 15 min (4°C). Remove the supernatant with a fine-tipped Pasteur pipette and transfer to a dimethyl sulphate waste bottle containing 5 M sodium hydroxide.

(vi) Gently rinse the wall of the Eppendorf tube with 1 ml ice-cold 70% ethanol. Chill the tube at −70°C for 10 min and centrifuge for 5 min, and remove the supernatant.

(vii) Add 1 ml 95% ethanol.

(viii) Centrifuge for 5 min and remove the supernatant. Place under vacuum for a few minutes (desiccator).

(ix) Add 100 µl of 1 M piperidine (freshly diluted).

(x) Close the cap and dissolve the DNA (Vortex). Centrifuge for a few seconds (Eppendorf). Heat at 90°C for 30 min (place a heavy weight on top of the tubes in a water bath). Punch holes in the tube cap with a dissecting needle. Vacuum dry in a Buchler Vortex Evaporator. Redissolve the oligonucleotides in 15 µl of water, vacuum dry. Repeat the addition of 15 µl of water twice and vacuum dry.

(xi) Add 6 µl of gel loading dye (80% formamide, 20% sucrose, 0.3% bromophenol blue, 0.03% xylene cyanol).

(xii) Close the cap and redissolve the DNA (Vortex and manual agitation). Centrifuge for a few seconds (Eppendorf). Count 1 µl. Add more loading dye to give about 10 000 c.p.m./µl. Mix, centrifuge for a few seconds. Heat at 90°C for 1 min and quickly chill in ice water. Immediately load 1 µl per slot on a 20% sequencing gel (denaturing polyacrylamide gel; Section 5.2).

5.1.2 *Limited Oligonucleotide Cleavage at Guanines and Adenines (G + A)*

(i) Mix 5 µl of [^{32}P]oligonucleotide containing 1.5 µg of calf DNA with 10 µl of 3% diphenylamine and 1.5 mM EDTA in 98% formic acid.

(ii) Close the cap and mix well. Incubate at 37°C for 15 min. Chill on ice, add 45 µl of water, extract with ether (500 µl) four times. Dry the aqueous phase (lower phase) in a Buchler Vortex Evaporator.

(iii) Add 100 µl of 1.0 M piperidine (freshly diluted), and follow through the same procedure as described above for the guanine cleavage.

Purification and Sequence Analysis

5.1.3 *Limited Oligonucleotide Cleavage at Cytosines and Thymines (C + T)*

(i) Mix 5 μl of [^{32}P]oligonucleotide containing 1.5 μg of calf DNA with 15 μl of distilled water and 30 μl of hydrazine (95%) reagent grade.
(ii) Close the cap of the 1.5 ml Eppendorf tube and mix gently (manual agitation).
(iii) Incubate at 45°C for 20 min. Chill the tube to 0°C.
(iv) Add 200 μl of stop solution (0.3 M ammonium acetate, 0.1 mM EDTA) and 5 μl of *E. coli* tRNA (10 mg/ml).
(v) Add 750 μl of 95% ethanol.
(vi) Close the cap and mix well (invert the tube four times). The oligonucleotides are precipitated with ethanol and treated with piperidine as described above for the guanine cleavage.

5.1.4 *Limited Oligonucleotide Cleavage at Cytosines (C)*

(i) Mix 5 μl of [^{32}P]oligonucleotide containing 1.5 μg of calf DNA with 15 μl of 5 M NaCl and 30 μl of hydrazine (95%).
(ii) Close the cap of the 1.5 ml Eppendorf tube and mix gently (manual agitation). Incubate at 45°C for 20 min. Chill the tube to 0°C.
(iii) Add 200 μl of stop solution, 5 μl of *E. coli* tRNA (10 mg/ml) and 750 μl of 95% ethanol.
(iv) Oligonucleotides are precipitated with ethanol and then treated with piperidine as described for the guanine cleavage.

5.1.5 *Limited Oligonucleotide Cleavage at Adenines and Cytosines (A > C)*

(i) Mix 5 μl of [^{32}P]oligonucleotide containing 1.5 μg of calf DNA with 100 μl of 1.2 M NaOH-1 mM EDTA, and heat at 90°C for 15 min (put a weight on top of the tube).
(ii) Add 150 μl of 1 N acetic acid, 5 μl of tRNA (50 μg total) and 750 μl of 95% ethanol.
(iii) Chill at −70°C for 30 min. Centrifuge for 15 min at 4°C.
(iv) Add 1 ml of 95% ethanol and chill at −70°C for 15 min, centrifuge for 10 min.
(v) Add 1 ml of 75% ethanol.
(vi) Centrifuge for 5 min. Place the tube under vacuum for a few minutes (desiccator), treat the oligonucleotides with piperidine as described above for the guanine cleavage reaction.

5.1.6 *Limited Oligonucleotide Cleavage at Thymines (T)*

(i) Use 5 μl of [^{32}P]oligonucleotide containing 1.5 μg of calf DNA, chill for 5 min.
(ii) Add 10 μl of 5% osmium tetraoxide (0°C) and 1 μl of pyridine (0°C).
(iii) Close the cap of the 1.5 ml Eppendorf tube and mix the contents well. Incubate at 0°C for 3 min.

(iv)   Add 200 µl 0.3 M NaAc (pH 5.2) containing 50 µg tRNA, and 700 µl 95% ethanol, mix gently. Keep at −70°C in a dry ice-isopropanol bath for 30 min.
(v)   Transfer the tube from −70°C bath to room temperature. When sample warms up and becomes fluid, centrifuge for 15 min at 4°C in an Eppendorf centrifuge. Remove supernatant.
(vi)   Add 200 µl of 0.3 M NaAc (pH 5.2). Mix, keep in an ice bath for 5 min. Add 700 µl of cold 95% ethanol, mix gently. Keep at −70°C for 10 min.
(vii) Centrifuge for 5 min at 4°C.
(viii) Repeat the ethanol precipitation and centrifugation once more. Remove the supernatant. Place the tube under vacuum for a few minutes (desiccator or Buchler Vortex Evaporator).
(ix)   Add 100 µl of 1 M piperidine (freshly diluted), and follow through the same procedure as described above for the guanine cleavage.

5.2 Sequencing Gels

In sequencing oligodeoxyribonucleotides by the limited chemical cleavage technique, the final step is gel electrophoresis to fractionate the nested sets of oligonucleotides by size. Labelled oligonucleotide fragments in the partially degraded set have one end in common (the 5' end in the case of the polynucleotide kinase reaction) and another end which varies in length. When electrophoresed through polyacrylamide gel, the larger fragment moves slower than the smaller one. Banaszuk et al. (11) used 24.8% gels with dimensions of 17 x 16 x 0.075 cm to fractionate oligonucleotides up to 14 residues in length. In our experience for fractionation of oligonucleotides 10–30 residues in length, 20% gels with dimensions of 40 x 34 (or 20) x 0.05 cm are more suitable. Sequencing gels have the composition of 19% acrylamide + 1% bisacrylamide containing 7 M urea, 80 mM Tris-borate (pH 8.3), 2 mM EDTA, 0.07% ammonium persulphate, and 10–50 µl N,N,N',N'-tetramethylethylenediamine (TEMED) so that polymerisation occurs within 10–15 min. We have found that a 20% gel is easier to handle than a 24.8% gel. Also a running buffer of 80 mM (or 100 mM) Tris-borate (pH 8.3), 2 mM EDTA is better than 50 mM Tris-borate. Details for preparing the gels and diagnosis of aberrations in sequencing band patterns can be found in the excellent article by Maxam and Gilbert (10) (see also Chapter 3).

A simplified method to form slots in preparation of a sequencing gel was perfected by Robert Yang in our laboratory 5 years ago.

(i)   Cut a slot-former from a piece of 30 x 6 x 0.035 cm Mylar (Almac Plastics, 1640 Emerson, Rochester, NY, USA) (the same material used for spacers), using a pair of sharp scissors to give a saw-toothed pattern (*Figure 2*).
(ii) After pouring the polyacrylamide mixture into the gel mould, place the flat side of the slot-former on the top surface.
(iii) After the polymerisation reaction is completed, the slot-former can be easily removed from the gel mould.

Purification and Sequence Analysis

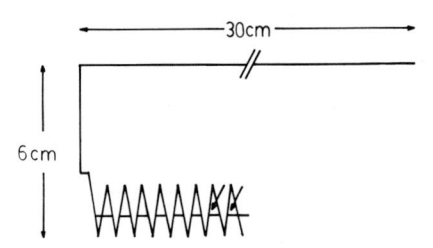

Figure 2. A slot-former for a sequencing gel. The tips of the slot-former are 6 mm apart. After pushing the tips into the gel by about 7 mm (the part below the dotted line is inside the gel), the width of each slot becomes approximately 3–4 mm. A sample of 1 μl is pipetted into each slot (see position of the short arrows) with the slot-former fixed in place.

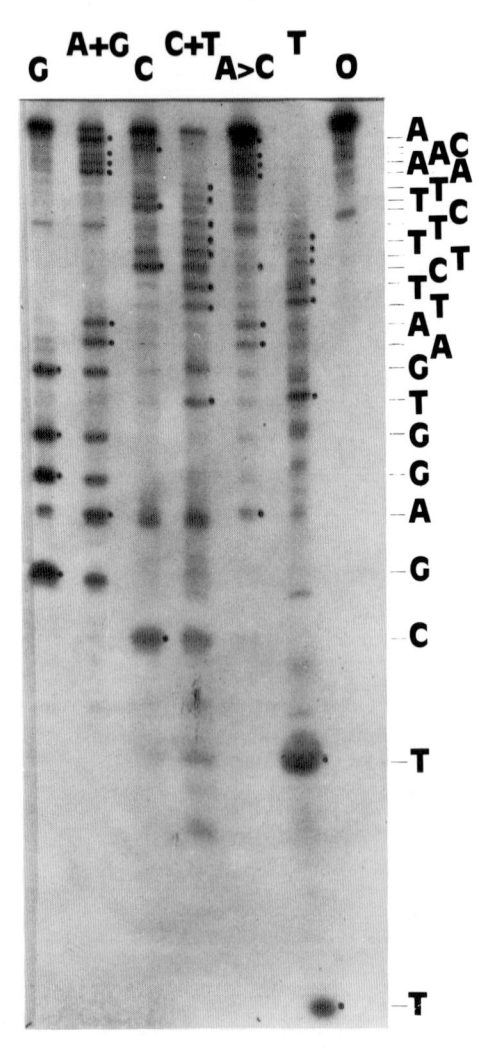

Figure 3. Sequence analysis of 5′ d(T-T-C-G-A-G-G-T-G-A-A-T-T-C-T-T-T-C-T-T-A-A-A-C-A) by the chemical cleavage method. 20% polyacrylamide gel, 1500 V, 1 h 45 min.

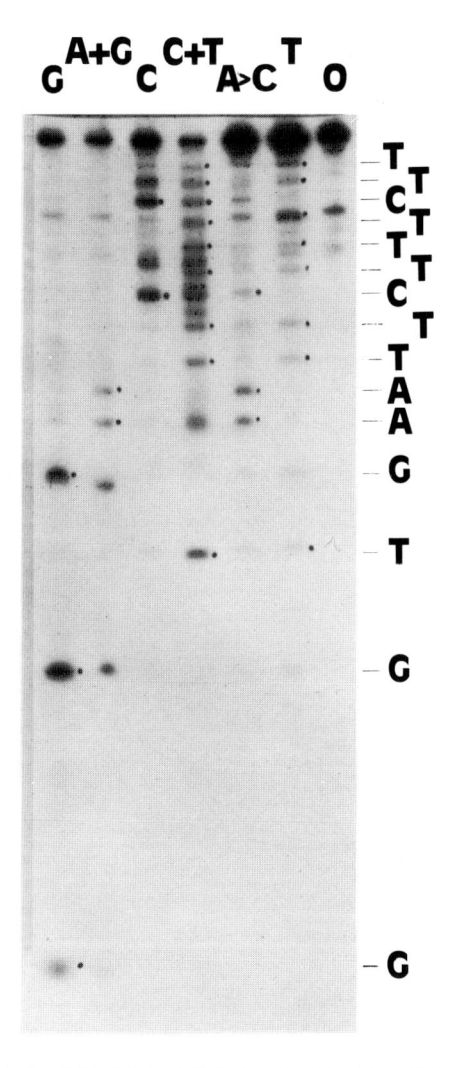

Figure 4. Sequence analysis of 5′ d(G-G-T-G-A-A-T-T-C-T-T-T-C-T-T) by the chemical cleavage method.

(iv) Turn it 180° and insert it on top of the flat gel surface so that the tip of the slot-former is about 7 mm inside the gel. The slot-former stays on top of the gel during the entire electrophoresis procedure.

(v) Load the samples in the slots in between saw-tooths. In order to get sharp bands it is best to load a 1 µl sample (containing ~10 000 c.p.m.) for a slot approximately 3–4 mm wide.

To read the sequence from the autoradiograph of a sequencing gel start from the lowest band. When only four standard cleavage reactions (G, A+G, T+C, C) are used, it is not always possible to read the entire sequence unambiguously. For example, in *Figure 3*, there are several places where a T cannot be distinguish-

Purification and Sequence Analysis

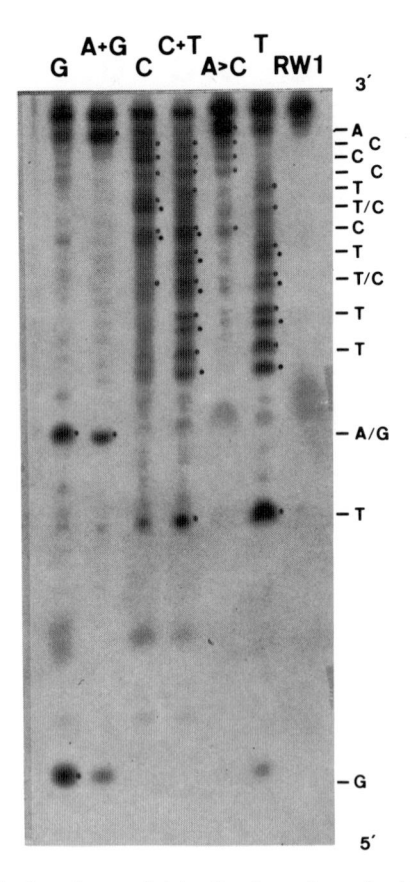

Figure 5. Sequence analysis of a mixture of eight oligodeoxyribonucleotides by the chemical cleavage method.

ed from a C. Adding a T-specific reaction (13) and a A > C reaction (10) usually allows an unambiguous reading of the sequence (*Figure 4*). Sometimes the first one or two nucleotides from the labelled end are weak or not very clear. It may be necessary to use the same end-labelled oligonucleotide and mobility-shift method (Section 3) to resolve the ambiguity. The first few nucleotides from the labelled end of an oligonucleotide are the easiest to read by the mobility-shift method.

When oligonucleotides are required as hybridisation probes, it is often necessary to synthesise several related oligonucleotides to include the degeneracy of the genetic code. Here the chemical cleavage method for sequencing oligonucleotides can also be used to determine the sequence of a mixture of oligonucleotides sharing the same 5' end. For example, a mixture of eight oligonucleotides has been synthesised to identify the rice cytochrome c gene. The mixture is composed of:

$$5' \text{ GT}^A_G \text{ TT}^T_C \text{ TC}^T_C \text{ TCC CA}$$

which represents the following eight compounds:

5' GTA TTT TCT TCC CA
5' GTG TTT TCT TCC CA
5' GTA TTC TCT TCC CA
5' GTG TTC TCT TCC CA
5' GTA TTT TCC TCC CA
5' GTG TTT TCC TCC CA
5' GTA TTC TCC TCC CA
5' GTG TTC TCC TCC CA

After phosphorylating the 5' ends of the above compounds with ^{32}P, the sequences in the mixture can be seen in *Figure 5*. There is one band for the first (G) or the second position (T). However, there are two closely spaced bands for the third position which represents 5' GTA and GTG, respectively. Since the electrophoretic mobility on a polyacrylamide gel depends on the base composition, these two trinucleotides move to slightly different positions. From the fifth position onwards, there are two or more closely-spaced bands for each position reflecting the nature of the mixture.

6. REFERENCES

1. Narang,S.A., Hsiung,H.M. and Brousseau,R. (1979) *Methods Enzymol.*, **68**, 90.
2. Mateucci,M.D. and Caruthers,M.H. (1981) *J. Am. Chem. Soc.*, **103**, 3186.
3. Wu,R., Bahl,C.P. and Narang,S.A. (1978) in *Progress in Nucleic Acid Research and Molecular Biology*, Vol. 21, Cohn,W.E. (ed.), Academic Press, p. 101.
4. Brownlee,G.T. and Sanger,F. (1969) *Eur. J. Biochem.*, **11**, 395.
5. Ling,V. (1972) *J. Mol. Biol.*, **64**, 87.
6. Sanger,F., Donelson,J.E., Coulson,A.D., Kössel,H. and Fischer,D. (1973) *Proc. Natl. Acad. Sci. USA*, **70**, 1209.
7. Wu,R., Tu,C.D. and Padmanabhan,R. (1973) *Biochem. Biophys. Res. Commun.*, **55**, 1092.
8. Tu,C.D., Jay,E., Bahl,C.P. and Wu,R. (1976) *Anal. Biochem.*, **74**, 73.
9. Tu,C.D. and Cohen,S.N. (1980) *Gene*, **10**, 177.
10. Maxam,A.M. and Gilbert,W. (1980) *Methods Enzymol.*, **65**, 499.
11. Banaszuk,A.M., Deugau,K.V., Sherwood,J., Michalak,M. and Glick,B.R. (1983) *Anal. Biochem.*, **128**, 281.
12. Rubin,C.M. and Schmid,C.W. (1981) *Nucleic Acids Res.*, **8**, 4613.
13. Friedmann,T. and Brown,D.M. (1978) *Nucleic Acids Res.*, **5**, 615.
14. Wu,R., Jay,E. and Roychoudhury,R. (1976) *Methods Cancer Res.*, **12**, 88.
15. Hindley,J. (1983) *Laboratory Techniques in Biochemistry and Molecular Biology: DNA Sequencing,* published by Elsevier Biomedical Press.
16. Griffin,B.E. (1971) *FEBS Lett.*, **15**, 165.
17. Michniewicz,J.J., Bahl,C.P., Itakura,K., Katagiri,N. and Narang,S.A. (1973) *J. Chromatogr.*, **85**, 159.
18. Jay,E., Bambara,R., Padmanabhan,R. and Wu,R. (1974) *Nucleic Acids Res.*, **1**, 331.
19. Roychoudhury,R., Fischer,D. and Kössel,H. (1971) *Biochem. Biophys. Res. Commun.*, **45**, 430.

CHAPTER 7

Chemical Synthesis of Small Oligoribonucleotides in Solution

JACQUES H. VAN BOOM and CARL T.J. WREESMANN

1. INTRODUCTION

The synthesis of RNA fragments by a phosphotriester approach is more elaborate and time-consuming than the preparation of DNA fragments (see Chapters 3 and 4). The reason for this is that ribonucleosides contain an additional 2′-hydroxyl (2′-OH) function. Thus, in the synthesis of RNA, three hydroxyl groups must be dealt with: one primary (5′-OH) and two secondary hydroxyl groups (2′-OH and 3′-OH) which have a *cis* configuration. In the last decade different approaches have been developed (1,2) for the synthesis of protected ribonucleosides which may serve as units suitable for the synthesis of RNA fragments by a phosphotriester approach. In this chapter we present the synthesis of two types of building units: terminal and non-terminal units. The terminal units (compounds **2a–e** in *Figure 1*) have a free 3′-hydroxyl group. The other two hydroxyl groups are protected with different acid-labile groups: the 5′-OH with a lipophilic trityl derivative (3) (C_{16}-DMTr) and the 2′-OH with the tetrahydropyranyl (4) (THP) group.

On the other hand the non-terminal units (compounds **1a–e** in *Figure 1*) carry unprotected 3′-OH and 5′-OH groups, while the 2′-OH is protected with the THP group. These protective groups have been chosen because the C_{16}-DMTr group is easily accessible and lipophilic which simplifies purification of the intermediate RNA phosphotriesters. Furthermore, the C_{16}-DMTr group can be removed under acidic conditions which do not lead to the isomerisation of the required 3′,5′-phosphodiester linkages. The use of the THP group is based on the fact that 2,3-dihydropyran is a commercially available and cheap reagent for the introduction of this group, which can also be removed selectively without any side reactions. The only disadvantage of the application of the THP group is that it introduces a chiral centre. The latter requires an additional purification step by column chromatography. The synthesis of the previously mentioned two types of units starts with the preparation of the non-terminal units **1a–e**. Key intermediates **1a–e** are prepared from 3′,5′-protected ribonucleosides which are treated with 2,3-dihydropyran, in the presence of acid, followed by the removal of the 3′,5′-protecting groups. The precursor of 2′-O-tetrahydropyranyluridine is obtained from 3′,5′-di-O-acetyl-uridine (5). The synthesis of the other intermediate 3′,5′-protected ribonucleosides is accomplished by using the bifunctional reagent 1,3-dichloro-1,1,3,3-tetra-isopropyldisiloxane (6) which can be prepared in large quantities.

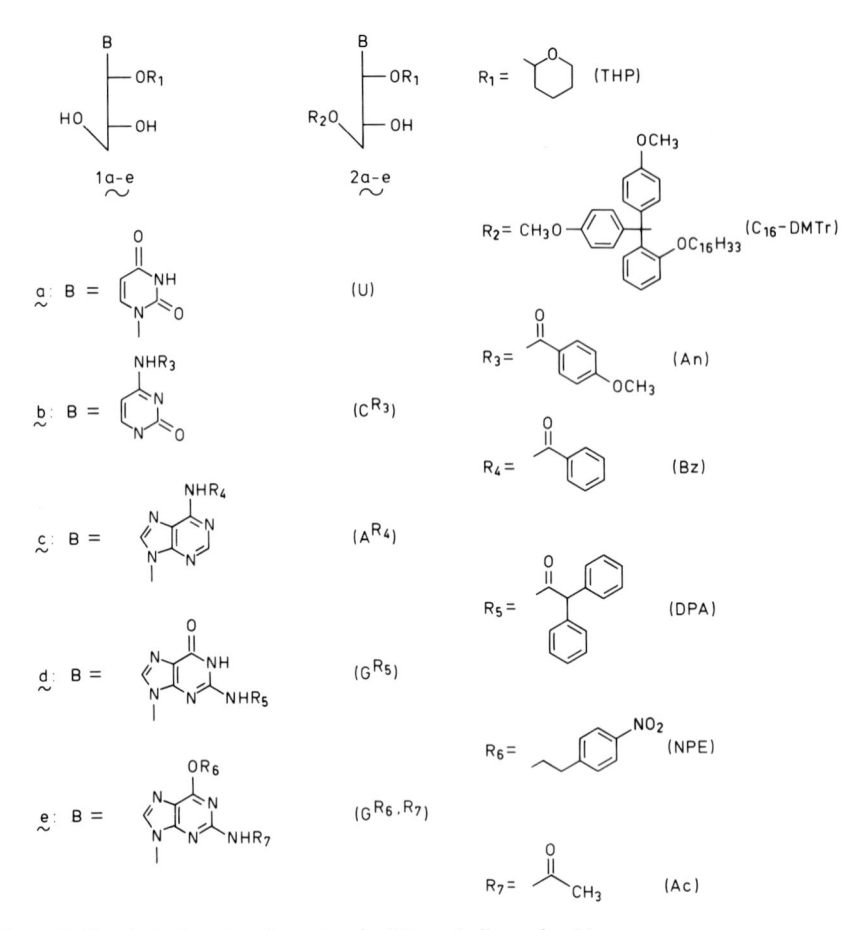

Chemical Synthesis of Small Oligoribonucleotides

Figure 1. Terminal (**2a–e**) and non-terminal (**1a–e**) ribonucleosides.

The exocyclic amino functions of the three ribonucleosides, cytidine, adenosine and guanosine, are protected with the base-labile 4-methoxybenzoyl (An), benzoyl (Bz) and diphenylacetyl (DPA) groups, respectively (see Chapter 2).

We also prepared a terminal and a non-terminal unit (i.e., **2e** and **1e**, respectively) of guanosine in which the exocyclic amino functions are protected with an acetyl (Ac) group and the O-6-positions with the 4-nitrophenylethyl (NPE) group (7). The latter guanosine derivatives are more lipophilic than the corresponding guanosine derivatives **2d** and **1d**. The application of the 'double-protected' guanosine derivatives **1e** and **2e** is recommended for the synthesis in solution of guanosine-rich RNA fragments by a phosphotriester approach.

The stepwise introduction of intermediate 3′,5′-phosphotriester linkages between one terminal and five non-terminal units is demonstrated by the synthesis of the hexamer UAGGAU (*Figure 2*). In this approach the easily accessible bifunctional phosphorylating agent (**3**) plays a pivotal role (8). Thus phosphorylation of the terminal uridine derivative **2a** with **3** gives intermediate **4** which reacts, in the

154

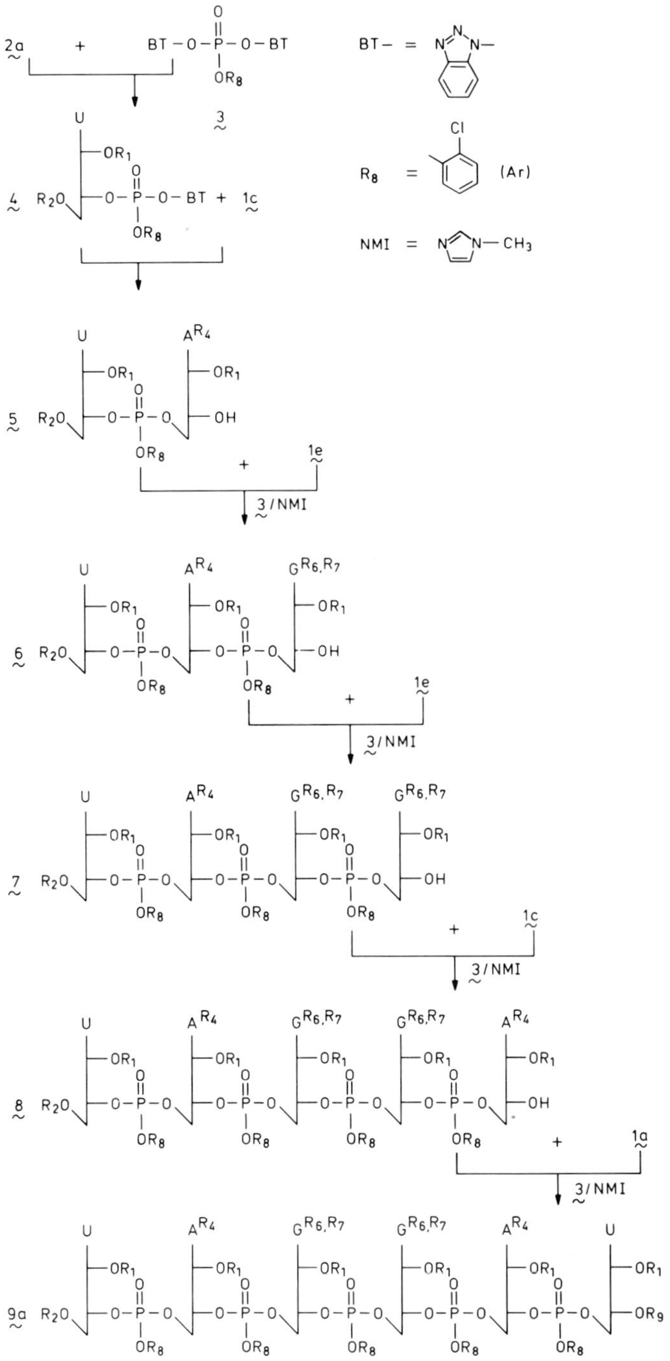

Figure 2. Assembly of the partially protected hexamer **9a** by a stepwise phosphorylation, using reagent **3**, from one terminal and five non-terminal units.

Chemical Synthesis of Small Oligoribonucleotides

Table 1. Unblocking of Hexamer **9a** to produce UAGGAU (**10**).

Compounds	Conditions as applied for the conversion of **9a** into **10**	Protecting groups[a]						
		R_1	R_2	R_4	R_6	R_7	R_8	R_9
9a		THP	C_{16}-DMTr	Bz	NPE	Ac	Ar	H
	toluene-*p*-sulphonic acid (19)							
9b		THP	H	Bz	NPE	Ac	Ar	H
	acetic anhydride							
9c		THP	Ac	Bz	NPE	Ac	Ar	Ac
	1° oximate (20, 21) 2° ammonia/H$_2$O 3° DBU (7)							
9d		THP	H	H	H	H	H	H
	0.01 N HCl							
10		H	H	H	H	H	H	H

[a]For the meaning of abbreviations used see *Figures 1* and *2*.

presence of N-methylimidazole, with the non-terminal unit **1c**, to produce (9) dimer **5** which has a free 3′-hydroxyl group.

Phosphorylation of dimer **5** with **3**, followed by the addition of the non-terminal unit **1e**, gives trimer **6**. Repetition of the above phosphorylation procedure three times and the addition of the appropriate non-terminal units, produces the partially protected hexamer **9a**. The latter is completely deblocked, as shown in *Table 1* to give UAGGAU (**10**).

The phosphotriester approach presented in this chapter differs in several aspects from that described in Chapter 4. In our approach an intermediate triester (e.g., **4**), which is activated by N-methylimidazole, reacts selectively with an incoming ribonucleoside which has two unprotected (3′- and 5′-) hydroxyl groups.

2. SYNTHESIS OF 2′-O-TETRAHYDROPYRANYLRIBONUCLEOSIDE MONOMERS

2.1 Synthesis of 2′-O-Tetrahydropyranyluridine (1a)

2.1.1 Reagents and Solvents

Uridine (Waldhof, FRG) is dried before use for 2 h at 50°C over phosphorus pentoxide *in vacuo*.

Dioxan, pyridine, trimethyl orthoacetate (Eastman Kodak) and 2,3-dihydropyran (Janssen) are dried by heating under reflux with calcium hydride (5 g/l) for 2 h and distilled. Toluene-*p*-sulphonic acid monohydrate (Baker), acetic acid (analytical grade, Baker) and acetic anhydride (analytical grade, Baker) are used without further purification.

Methanolic ammonia: dry (potassium hydroxide) ammonia gas is passed through methanol (500 ml, analytical grade, Baker) at −20°C until saturation and diluted with methanol (500 ml, analytical grade).

Melting points are uncorrected. T.l.c. is performed on silica gel (Schleicher and

Schüll TLC-Ready Plastic Sheets F1500 LS 254). Methylene chloride/methanol (92:8) (A), and methylene chloride/methanol (86:14) (B) are the solvent systems used. Sugar-containing material can be seen on t.l.c. plates by spraying with concentrated sulphuric acid/methanol (1:4) and charring at 150°C. Column chromatography is performed on silica gel (Merck, Kieselgel 7729). ^{1}H-n.m.r. spectra are measured at 100 MHz using a JEOL JNMPS 100 spectrometer. ^{13}C-n.m.r. spectra are measured at 25.15 MHz using a JEOL JNMPFT 100 spectrometer. Chemical shifts are given in p.p.m. (δ) relative to tetramethylsilane (TMS) as internal standard.

2.1.2 *3',5'-Di-O-acetyluridine* (5)

(i) Stir a mixture of dry uridine (24.4 g; 100 mmol), toluene-*p*-sulphonic acid monohydrate (100 mg), dry dioxan (150 ml) and dry trimethyl orthoacetate (17 ml) for 1 h at room temperature. T.l.c. (system B) should show complete conversion of the starting material.

(ii) Add a mixture of dry pyridine (150 ml) and acetic anhydride (50 ml) dropwise over 15 min at room temperature to the stirred reaction mixture and leave it overnight. T.l.c. (system B) should show that the acetylation has gone to completion.

(iii) Add methanol (50 ml) dropwise to the reaction mixture at 0°C to destroy excess acetic anhydride.

(iv) Concentrate the mixture under reduced pressure to a colourless oil, dissolve in methylene chloride (250 ml) and wash with an aqueous solution of sodium bicarbonate (1 M; 100 ml).

(v) Extract the water layer with methylene chloride (5 x 100 ml), dry the combined organic layers over magnesium sulphate (25 g) and concentrate to a small volume.

(vi) Dissolve in a mixture of glacial acetic acid and water (180 ml; 9:1). After 15 min at room temperature t.l.c. (system B) should show that the cleavage of the methoxyethylidene function is complete.

(vii) Evaporate the aqueous acetic acid under reduced pressure and co-evaporate the residue with toluene (2 x 100 ml) and ethanol (2 x 100 ml).

(viii) Keep the resulting white glass *in vacuo* in a desiccator over potassium hydroxide pellets until the smell of acetic acid can no longer be detected.

(ix) Dissolve the crude product in boiling ethanol (500 – 1000 ml) and keep the solution at room temperature.

(x) After 2 days at room temperature filter off the crystals. The isomeric purity should be ascertained with 100 MHz n.m.r. spectroscopy (10).

Yield 23.0 g (70% starting from uridine), m.p. 152 – 155°C, R_f 0.29 (system A), 0.60 (system B), ^{1}H-n.m.r. (C_2D_6SO/D_2O, 1:1, containing three drops of acetic acid per ml): 7.57 (H_6, d, J 7.5 Hz), 5.89 (H_5, d, J 7.5 Hz), 5.88 (H_1', d, J 5.5 Hz). The 2',5'-isomer shows a slightly different ^{1}H n.m.r. spectrum (C_2D_6SO/D_2O, 1:1, containing three drops of acetic acid per ml): 7.57 (H_6, d, J 7.5 Hz), 5.89 (H_5, d, J 7.5 Hz), 5.86 (H_1', d, J 6.0 Hz).

2.1.3 2'-O-Tetrahydropyranyluridine (1a) (4)

(i) Add 3',5'-di-O-acetyluridine (32.8 g; 100 mmol) and dry 2,3-dihydropyran (50 ml) to a clear and stirred solution of toluene-*p*-sulphonic acid monohydrate (138 mg) and dry dioxan (250 ml).

(ii) After 10 min at room temperature cool the clear solution in an ice-water bath. T.l.c. (system B) should indicate complete conversion of the starting material.

(iii) Neutralise the reaction mixture with methanolic ammonia, concentrate under reduced pressure and dissolve in methanolic ammonia (500 ml). After standing overnight, t.l.c. (system B) should show complete deacetylation.

(iv) Evaporate off the methanolic ammonia and apply the residue to a column of silica gel (300 g; 12 x 7 cm) packed in methylene chloride.

(v) Elute the column with methylene chloride/methanol (500 ml, 98:2; 500 ml, 96:4; 1000 ml, 9:1; 1000 ml, 8:1).

(vi) Collect the fractions containing both diastereoisomers and re-chromatograph on a column of silica gel (300 g; 12 x 7 cm) packed in methylene chloride.

(vii) Elute the column with methylene chloride/methanol (500 ml, 98:2; 500 ml, 96:4; 1000 ml, 9:1; 1000 ml, 8:2) and collect the fractions containing both diastereoisomers.

(viii) Concentrate to a small volume and apply to another column of silica gel (150 g; 12 x 3.5 cm) packed in methylene chloride.

(ix) Elute the column with methylene chloride/methanol (200 ml, 98:2; 300 ml, 96:4; 500 ml, 8:2).

(x) Collect the fractions containing one of two diastereoisomers in its pure form and concentrate to a colourless oil.

(xi) Crystallise the high running diastereoisomer of 2'-O-tetrahydropyranyluridine from ethyl acetate. This should yield 13.6 g (41%), m.p. 146–148°C, R_f 0.28 (system A), 0.59 (system B), ^{1}H-n.m.r. (CDCl$_3$/CD$_3$OD): 7.83 (H$_6$, d, J 7.5 Hz), 5.83 (H$_1'$, d, J 3.0 Hz), 5.60 (H$_5$, d, J 7.5 Hz). ^{13}C-n.m.r. (CDCl$_3$/CD$_3$OD): 165.1 (C$_4$), 151.1 (C$_2$), 141.4 (C$_6$), 102.7 (C$_5$), 88.8 (C$_1'$), 70.7 (C$_4'$), 69.5 (C$_3'$), 64.7 (C$_2'$), 60.9 (C$_5'$), 100.3, 85.1, 31.1, 25.4, 20.6 (THP).

(xii) Crystallise the low-running diastereoisomer from ethanol: yield 13.8 g (42%), m.p. 185–187°C, R_f 0.20 (system A), 0.48 (system B), ^{1}H-n.m.r. (CDCl$_3$/CD$_3$OD/C$_2$D$_6$SO): 7.73 (H$_6$, d, J 7.5 Hz), 5.95 (H$_1'$, d, J 4.5 Hz), 5.56 (H$_5$, d, J 7.5 Hz). ^{13}C-n.m.r. (CDCl$_3$/CD$_3$OD/C$_2$D$_6$SO): 165.0 (C$_4$), 151.7 (C$_2$), 142.2 (C$_6$), 102.7 (C$_5$), 88.0 (C$_1'$), 78.8 (C$_4'$), 70.1 (C$_3'$), 62.6 (C$_2'$), 62.1 (C$_5'$), 98.8, 86.3, 30.7, 26.0, 19.7 (THP).

2.2 Synthesis of 2'-O-Tetrahydropyranyl-4-N-anisoylcytidine (1b)

2.2.1 Reagents and Solvents

Dry cytidine (Waldhof, FRG) before use for 2 h at 50°C over phosphorus pentoxide *in vacuo*. Dry N,N-dimethylformamide by stirring overnight at room temperature with calcium hydride (5 g/l) and by distilling under reduced pressure (b.p. 70 – 80°C, 20 – 30 mm Hg). Dry acetonitrile and triethylamine by heating under reflux with calcium hydride (5 g/l) followed by distillation. For dioxan, pyridine and 2,3-dihydropyran see Section 2.1.1. For methanolic ammonia see Section 2.1.1. Dry 1-hydroxybenzotriazole (Fluka) before use for 72 h at 50°C over phosphorus pentoxide *in vacuo*. **(WARNING: at higher temperature 1-hydroxybenzotriazole may explode).** Potassium fluoride (Baker) and tetraethylammonium bromide (Fluka) are used without further purification.

Triethylammonium bicarbonate buffer (2 M). Saturate a mixture of triethylamine (825 ml) and water (2175 ml) with carbon dioxide gas at 0°C until the pH of the clear solution becomes 7.5.

Mesitylenesulphonic acid.
(i) Shake a mixture of mesitylene (100 ml) and concentrated sulphuric acid (36 M; 200 ml) vigorously for 10 min at room temperature.
(ii) Pour the mixture into crushed ice.
(iii) Filter off the crystals formed, wash with concentrated hydrochloric acid (13.6 N; 200 ml, 0°C) and dry over potassium hydroxide pellets *in vacuo*.
(iv) Recrystallise the crude product from chloroform (200 ml, analytical grade, Baker) and leave at – 20°C.
(v) Filter off the crystals and dry for 2 weeks at room temperature over concentrated sulphuric acid *in vacuo*.
(vi) Store mesitylenesulphonic acid over concentrated sulphuric acid at 0°C *in vacuo*.

Anisoyl chloride.
(i) Heat a mixture of anisoic acid (152 g, Janssen) thionyl chloride (120 ml, Merck) and dry dimethylformamide (1.0 ml) under reflux for 1 h.
(ii) Evaporate the excess thionyl chloride under reduced pressure and obtain pure anisoyl chloride by distillation (b.p. 155 – 160°C, 2 – 3 mm Hg). Yield 130 g.
(iii) Store anisoyl chloride in a dark bottle at 0°C.

1,3-Dichloro-1,1,3,3-tetraisopropyldisiloxane.
(i) Add to a mixture of magnesium curls (64 g) and dry ether (200 ml, distilled from phosphorus pentoxide, 30 g/l) dropwise a solution of isopropyl bromide [270 ml, Janssen, distilled from calcium hydride (5 g/l) before use] in dry ether (400 ml).
(ii) Stir the reaction mixture mechanically and heat under reflux for 3.5 h and then add a solution of trichlorosilane (100 ml, Janssen, distilled before use) and dry ether (400 ml) dropwise to the stirred solution.
(iii) Heat the stirred reaction mixture under reflux overnight and quench by adding dropwise hydrochloric acid (0.1 N; 800 ml).
(iv) Stir the mixture and heat under reflux for another 3.5 h.
(v) Separate the organic layer and extract the aqueous layer with ether (3 x 300 ml).
(vi) Dry the combined organic layers over magnesium sulphate and concentrate to a colourless oil under reduced pressure.

(vii) Distil the residue (b.p. 80–90°C, 10 mm Hg) to produce pure 1,1,3,3-tetraisopropyldisiloxane (100 g) as a colourless oil.
(viii) Dissolve the above product in methylene chloride (500 ml, analytical grade, Baker) and pass a stream of dry (H_2SO_4 dried) chlorine gas through the solution.
(ix) When the temperature rises to about 27–30°C cool the reaction mixture to 17–20°C by immersion in an ice-water bath, while the stream of chlorine gas is continued.
(x) After 2 h, and then after every hour, withdraw a small sample of the reaction mixture and analyse by i.r. spectroscopy (NaCl).
(xi) Stop the chlorination when the i.r. spectrum indicates the disappearance of the absorption band at 2100 cm^{-1} (Si-H).
(xii) Evaporate the volatile compounds and distil the residue under diminished pressure (b.p. 85–90°C, 2 mm Hg) to give 1,3-dichloro-1,1,3,3-tetraisopropyldisiloxane (70 g) as a colourless oil, and store at 0°C.

For melting points, t.l.c., column chromatography and n.m.r. spectroscopy, see Section 2.1.1.

2.2.2 3′,5′-O-(Tetraisopropyldisiloxane-1,3-diyl)cytidine (6)

(i) To a stirred suspension of dry cytidine (12.15 g; 50 mmol) in dry dimethylformamide (200 ml) and dry pyridine (40 ml) add dropwise over a period of 15 min a solution of 1,3-dichloro-1,1,3,3-tetraisopropyldisiloxane (18.0 ml; 62.5 mmol) in dry dimethylformamide (50 ml).
(ii) After 1 h at room temperature, t.l.c. (system B) should indicate complete conversion of the starting material.
(iii) Neutralise the reaction mixture with triethylammonium bicarbonate buffer (2 M; 75 ml), concentrate under reduced pressure to a small volume (100 ml), dissolve in methylene chloride (1000 ml) and wash with an aqueous solution of sodium bicarbonate (1 M; 500 ml) and water (500 ml).
(iv) Dry the organic layer over magnesium sulphate (50 g) and concentrate to a colourless oil.
(v) Crystallise from acetone (500 ml) to produce pure 3′,5′-O-(tetraisopropyldisiloxane-1,3-diyl)cytidine. Yield 19.5 g (85%), m.p. 226–228 (decomp.), R_f 0.22 (system A). ^{1}H-n.m.r. ($CDCl_3/CD_3OD$): 8.00 (H_6, d, J 7.5 Hz), 7.58 (H_5, d, J 7.5 Hz), 5.64 (H_1, s), ^{13}C-n.m.r. ($CDCl_3/CD_3OD$): 166.2 (C_4), 156.3 (C_2), 140.6 (C_6), 94.7 (C_5), 91.5 (C_1'), 81.6 (C_4'), 75.1 (C_3'), 68.2 (C_2'), 60.0 (C_5'), 17.4, 17.0, 13.5, 13.0, 12.5 (TIPS).

2.2.3 3′,5′-O-(Tetraisopropyldisiloxane-1,3-diyl)-4-N-anisoylcytidine

(i) To a stirred solution of dry 1-hydroxybenzotriazole (11) (10.0 g; 75 mmol) in dry dioxan (300 ml) and dry triethylamine (21 ml; 150 mmol) add dropwise a solution of anisoyl chloride (10.3 ml; 75 mmol) in dry dioxan (50 ml) over 15 min.

(ii) After stirring for 1 h, filter off the precipitated triethylammonium chloride salts and add the filtrate to a solution of 2'-O-(tetraisopropyldisiloxane-1,3-diyl)cytidine (24.25 g; 50 mmol) in dry dimethylformamide (150 ml).

(iii) Evaporate off a volume of about 100 ml under reduced pressure and stir the residue at room temperature for 3 days. T.l.c. (system A) should indicate complete conversion of the starting material.

(iv) Add water (5 ml) and concentrate the reaction mixture under reduced pressure to a small volume (100 ml), dissolve in methylene chloride (1000 ml) and wash with an aqueous solution of sodium bicarbonate (1 M; 500 ml) and water (500 ml).

(v) Dry the organic layer over magnesium sulphate (50 g) and concentrate to a light brown oil.

(vi) Crystallise from acetonitrile to afford pure 3',5'-O-(tetraisopropyldisiloxane-1,3-diyl)-4-N-anisoylcytidine. Yield 29.6 g (90%), m.p. 117–120°C, R_f 0.58 (system A), ^{1}H-n.m.r. (CDCl$_3$): 8.40 (H_6, d, J 7.5 Hz), 7.95 (anisoyl, 2H, d, J 9.0 Hz), 7.50 (H_5, d, J 7.5 Hz), 6.95 (anisoyl, 2H, d, J 9.0 Hz), 5.84 (H_1', s), 3.80 (anisoyl, 3H, s), ^{13}C-n.m.r. (CDCl$_3$): 166.0 (C_4), 154.7 (C_2), 144.1 (C_6), 96.3 (C_5), 91.5 (C_1'), 81.8 (C_4'), 75.0 (C_3'), 68.4 (C_2'), 59.9 (C_5'), 163.4, 162.6, 126.7, 125.0, 114.0, 55.4 (anisoyl), 17.3, 17.2, 16.9, 13.3, 12.3 (TIPS).

2.2.4 *2'-O-Tetrahydropyranyl-4-N-anisoylcytidine* (1b)

(i) To a stirred suspension of 3',5'-O-(tetraisopropyldisiloxane-1,3-diyl)-4-N-anisoylcytidine (31.0 g; 50 mmol) in dry acetonitrile (250 ml) and dry dioxan (150 ml) add dry 2,3-dihydropyran (25 ml) and mesitylenesulphonic acid (3.0 g).

(ii) After stirring for 3 h at room temperature stir the clear solution overnight. T.l.c. (system A) should indicate complete conversion of the starting material.

(iii) Neutralise the reaction mixture with methanolic ammonia and concentrate under reduced pressure to a small volume (100 ml).

(iv) Dissolve in methylene chloride (1000 ml) and wash with an aqueous solution of sodium bicarbonate (1 M; 500 ml) and water (500 ml).

(v) Dry the organic layer over magnesium sulphate (50 g) and concentrate.

(vi) Dissolve the resulting oil, which contains both diastereoisomers of 2'-O-tetrahydropyranyl-3',5'-O-(tetraisopropyldisiloxane-1,3-diyl)-4-N-anisoylcytidine, in acetonitrile (500 ml).

(vii) To the solution add an aqueous solution of potassium fluoride (5 M; 50 ml) and tetraethylammonium bromide (50 g).

(viii) Stir the resulting mixture for 1 h at 60°C. T.l.c. (system A) should show complete removal of the silyl group.

(ix) Concentrate the reaction mixture under reduced pressure to a small volume (100 ml).

(x) Take up the residue in methylene chloride (1000 ml) and wash with an aqueous solution of sodium bicarbonate (1 M; 500 ml) and water (500 ml).

Chemical Synthesis of Small Oligoribonucleotides

(xi) Dry the organic layer over magnesium sulphate (50 g), concentrate to a small volume (100 ml) and triturate with a mixture of petroleum-ether (40–60°C) and ether (1000 ml, 1:1).

(xii) Filter off the precipitate and dissolve in methylene chloride.

(xiii) Evaporate off the solvent to give a colourless glass and crystallise from ethanol (500 ml) to produce the pure low-running diastereoisomer of 2'-O-tetrahydropyranyl-4-N-anisoylcytidine. Yield 6.5 g (28%), m.p. 182–184°C.

(xiv) Concentrate the mother liquor to a small volume and apply to a column of silica gel (300 g; 12 x 7 cm) packed in methylene chloride.

(xv) Elute the column with methylene chloride/methanol (500 ml, 98:2; 500 ml, 96:4; 1000 ml, 9:1).

(xvi) Collect the fractions containing both diastereoisomers and re-chromatograph on a column of silica gel (150 g; 12 x 3.5 cm) packed in methylene chloride.

(xvii) Collect the fractions containing one of the two diastereoisomers in its pure form and concentrate to a glass.

Yield of the high-running diastereoisomer of 2'-O-tetrahydropyranyl-4-N-anisoylcytidine is 7.2 g (32%), R_f 0.46 (system A), ^{1}H-n.m.r. (C_2D_6SO): 12.51 (N-H, broad s), 8.62 (H_6, d, J 7.5 Hz), 8.06 (anisoyl, 2H, d, J 9.0 Hz), 7.41 (H_5, d, J 7.5 Hz), 7.04 (anisoyl, 2H, d, J 9.0 Hz), 5.83 (H_1', s), 3.81 (anisoyl, 3H, s), ^{13}C-n.m.r. (C_2D_6SO): 163.1 (C_4), 154.3 (C_2), 144.6 (C_6), 96.2 (C_5), 88.9 (C_1'), 81.7 (C_4'), 71.9 (C_3'), 65.8 (C_2'), 62.8 (C_5'), 166.5, 162,8, 130.6, 126.2, 113.7, 55.5 (anisoyl), 96.1, 84.0, 30.2, 25.1, 18.4 (THP). Yield of the low-running diastereoisomer is 3.6 g (16%), R_f 0.35 (system A), ^{1}H-n.m.r. (C_2D_6SO): 12.53 (N-H, broad s), 8.50 (H_6, d, J 7.5 Hz), 8.11 (anisoyl, 2H, d, J 9.0 Hz), 7.41 (H_5, d, J 7.5 Hz), 7.06 (anisoyl, 2H, d, J 9.0 Hz), 6.09 (H_1', d, J 3.6 Hz), 3.83 (anisoyl, 3H, s), ^{13}C-n.m.r. (C_2D_6SO): 162.7 (C_4), 154.3 (C_2), 145.1 (C_6), 96.7 (C_5), 88.1 (C_1'), 77.6 (C_4'), 68.4 (C_3'), 60.7 (C_2'), 60.4 (C_5'), 166.0, 162,8, 130.5, 125.2, 113.5, 55.3 (anisoyl), 96.4, 85.1, 29.6, 24.8, 18.3 (THP).

2.3 Synthesis of 2'-O-Tetrahydropyranyl-6-N-benzoyladenosine (1c)

2.3.1 Reagents and Solvents

Dry adenosine (Waldhof, FRG) before use for 2 h at 50°C over phosphorus pentoxide *in vacuo*.

For dioxan, pyridine and 2,3-dihydropyran see Section 2.1.1. For dimethylformamide and acetonitrile see Section 2.2.1. For methanolic ammonia see Section 2.1.1. For triethylammonium bicarbonate buffer (2 M) see Section 2.2.1. For mesitylenesulphonic acid see Section 2.2.1. For 1,3-dichloro-1,1,3,3-tetraisopropyldisiloxane see Section 2.2.1. Trimethylsilyl chloride (Merck), benzoyl chloride (analytical grade, Baker) and concentrated ammonia (14.8 M, analytical grade, Baker) are used without further purification.

Solution of tetrabutylammonium fluoride in dioxan (1 M). Neutralise an aqueous solution of tetrabutylammonium hydroxide (1.6 M; 130 ml, Janssen) at 0°C with an aqueous solution of hydrofluoric acid (20 N; 10 ml, Baker) until the pH becomes 7–8 on wet pH-paper (Universal, Merck). Concentrate the solution to a small volume and co-evaporate with dry dioxan (3 x 200 ml). Add to the residue dry dioxan until the total volume is 200 ml, and store the solution (1 M) at room temperature.

For melting points, t.l.c., column chromatography and n.m.r. spectroscopy see Section 2.1.1.

2.3.2 3',5'-O-(Tetraisopropyldisiloxane-1,3-diyl)adenosine (6)

(i) To a stirred suspension of dry adenosine (13.35 g; 50 mmol) in dry dimethylformamide (200 ml) and dry pyridine (40 ml) add dropwise over a period of 15 min a solution of 1,3-dichloro-1,1,3,3-tetraisopropyldisiloxane (18.0 ml; 62.5 mmol) in dry dimethylformamide (50 ml).

(ii) After 1 h at room temperature, t.l.c. (system B) should indicate complete conversion of the starting material.

(iii) Neutralise the reaction mixture with triethylammonium bicarbonate buffer (2 M; 75 ml) concentrate under reduced pressure to a small volume (100 ml), dissolve in methylene chloride (1000 ml) and wash with an aqueous solution of sodium bicarbonate (1 M; 500 ml) and water (500 ml).

(iv) Dry the organic layer with magnesium sulphate (50 g) and concentrate to a colourless oil and crystallise from acetonitrile (500 ml) to afford pure 3',5'-O-(tetraisopropyldisiloxane-1,3-diyl)adenosine. Yield 20.5 g (80%), m.p. 135°C, R_f 0.35 (system A), ^{1}H-n.m.r. (CDCl$_3$): 8.32 (H$_8$, s), 8.10 (H$_2$, s), 6.06 (H$_1{'}$, s), ^{13}C-n.m.r. (CDCl$_3$): 155.9 (C$_6$), 152.8 (C$_2$), 148.4 (C$_4$), 138.9 (C$_8$), 119.9 (C$_5$), 90.3 (C$_1{'}$), 82.0 (C$_4{'}$), 75.1 (C$_3{'}$), 69.5 (C$_2{'}$), 60.8 (C$_5{'}$), 17.5, 17.4, 17.1, 16.9, 13.5, 13.2, 13.0, 12.6 (TIPS).

2.3.3 2'-O-Tetrahydropyranyl-3',5'-O-(tetraisopropyldisiloxane-1,3-diyl)adenosine

(i) To a stirred solution of 3',5'-O-(tetraisopropyldisiloxane-1,3-diyl)adenosine (25.5 g; 50 mmol) in dry dioxan (250 ml) add 2,3-dihydropyran (25 ml) and mesitylene sulphonic acid (12.5 g; 16.25 mmol). After 1 h at room temperature, t.l.c. (system A) should indicate complete conversion of the starting material.

(ii) Neutralise the reaction mixture with methanolic ammonia and concentrate under reduced pressure to a small volume (100 ml), dissolve in methylene chloride (1000 ml) and wash with an aqueous solution of sodium bicarbonate (1 M; 500 ml) and water (500 ml).

(iii) Dry the organic layer over magnesium sulphate (50 g) and concentrate to a light brown oil.

(iv) Crystallise the oil, which contains both diastereoisomers of 2'-O-tetrahydropyranyl-3',5'-O-(tetraisopropyldisiloxane-1,3-diyl)adenosine, from acetonitrile. Yield 23.8 g (80%), m.p. 172–174°C, R_f 0.46 (system A), ^{1}H-n.m.r. (CDCl$_3$, a mixture of two diastereoisomers): 8.32 (H$_8$, s), 8.29 (H$_8$, s), 8.17 (H$_2$, s), 8.04 (H$_2$, s), 6.06 (H$_1{'}$, d, J 6.0 Hz), 5.88 (H$_1{'}$, d, J 6.0 Hz).

2.3.4 2'-O-Tetrahydropyranyl-6-N-benzoyladenosine (12) (1c)

(i) To a stirred solution of 2'-O-tetrahydropyranyl-3',5'-O-(tetraisopropyl-disiloxane-1,3-diyl)adenosine (29.8 g; 50 mmol) in dry dioxan (125 ml) add a solution of tetrabutylammonium fluoride in dioxan (1 M; 125 ml). After 10 min at room temperature, t.l.c. (system B) should show complete removal of the silyl group.

(ii) Add water (10 ml) and concentrate the reaction mixture to a small volume (50 ml) and triturate with petroleum-ether (40–60°C, 1000 ml).

(iii) Remove the petroleum-ether by decantation and dilute the residue with methanol (20 ml), and apply to a column of silica gel (300 g; 12 x 7 cm) packed in methylene chloride.

(iv) Elute the column with methylene chloride/methanol (500 ml, 99:1; 500 ml, 98:2; 500 ml, 96:4; 1000 ml, 9:1; 1000 ml, 8:2).

(v) Collect the fractions containing both diastereoisomers and re-chromatograph on a column of silica gel (150 g; 12 x 3.5 cm) packed in methylene chloride.

(vi) Elute the column with methylene chloride/methanol (300 ml, 99:1; 300 ml, 98:2; 300 ml, 96:4; 500 ml, 9:1; 500 ml, 8:2).

(vii) Collect the fractions containing one of the two diastereoisomers in its pure form and concentrate to a glass. Yield of the high-running diastereoisomer of 2'-O-tetrahydropyranyladenosine is 5.6 g (32%), R_f 0.29 (system A). Yield of the low-running diastereoisomer is 6.7 g (38%). R_f 0.20 (system A).

(viii) Dissolve the high-running diastereoisomer of 2'-O-tetrahydropyranyladenosine (5.6 g; 16 mmol) in dry pyridine (80 ml) and add trimethylsilyl chloride (10.4 ml).

(ix) Stir the reaction mixture for 15 min at room temperature.

(x) Add benzoyl chloride (9.3 ml) and continue stirring for 2 h at room temperature.

(xi) Cool the reaction mixture to 0°C.

(xii) Add water (16 ml), and 5 min later, concentrated ammonia (14.8 M; 32 ml).

(xiii) After stirring for 30 min at room temperture, concentrate the reaction mixture under reduced pressure to give a colourless oil.

(xiv) Take up the oil in methylene chloride (800 ml) and wash with an aqueous solution of sodium bicarbonate (1 M; 150 ml) and water (150 ml).

(xv) Dry the organic layer over magnesium sulphate (20 g), concentrate to a small volume (15 ml) and triturate with petroleum-ether (40–60°C, 300 ml).

(xvi) Filter off the precipitate, dissolve in methylene chloride and concentrate to afford the pure high-running diastereoisomer of 2'-O-tetrahydropyranyl-6-N-benzoyladenosine as a colourless glass. Yield is 5.7 g (78%), R_f 0.49 (system A), ^{1}H-n.m.r. (CDCl$_3$/CD$_3$OD): 8.59 (H$_8$, s), 8.33 (H$_2$, s), 8.0–7.8 (benzoyl, 2H, m), 7.5–7.3 (benzoyl, 3H, m), 6.08 (H$_1{'}$, d, J 6.0 Hz), ^{13}C-n.m.r. (CDCl$_3$/CD$_3$OD): 152.2 (C$_6$), 151.0 (C$_2$), 150.6 (C$_4$), 143.4 (C$_8$),

124.1 (C$_5$), 89.2 (C$_1'$), 81.7 (C$_4'$), 71.9 (C$_3'$), 65.8 (C$_2'$), 62.8 (C$_5'$), 166.0, 133.6, 133.0, 128.8, 128.3 (benzoyl), 102.2, 87.3, 31.0, 25.0, 21.1 (THP).

Follow the same procedure for the low-running diastereoisomer. Yield (out of 5.6 g 2'-O-tetrahydropyranyladenosine) is 7.6 g (84%), R_f 0.37 (system A), ^{1}H-n.m.r. (CDCl$_3$/CD$_3$OD): 8.62 (H$_8$, s), 8.42 (H$_2$, s), 8.0 – 7.9 (benzoyl, m, 2H), 7.6 – 7.4 (benzoyl, m, 3H), 6.20 (H$_1'$, d, J 6.0 Hz), ^{13}C-n.m.r. (CDCl$_3$/CD$_3$OD): 152.3 (C$_6$), 152.0 (C$_2$), 151.7 (C$_4$), 144.4 (C$_8$), 124.1 (C$_5$), 89.3 (C$_1'$), 79.3 (C$_4'$), 70.7 (C$_3'$), 63.2 (C$_2'$), 62.7 (C$_5'$), 166.3, 134.0, 133.2, 129.0, 128.6 (benzoyl), 99.8, 87.6, 30.5, 30.5, 25.3, 19.8 (THP).

2.4 Synthesis of 2'-O-Tetrahydropyranyl-2-N-diphenylacetylguanosine (1d)

2.4.1 *Reagents and Solvents*

Dry guanosine (Waldhof, FRG) before use for 2 h at 50°C over phosphorus pentoxide *in vacuo*.

For dioxan, pyridine and 2,3-dihydropyran see Section 2.1.1. For dimethylformamide and acetonitrile see Section 2.2.1. For methanolic ammonia see Section 2.1.1. For mesitylene-sulphonic acid see Section 2.2.1. For 1,3-dichloro-1,1,3,3-tetraisopropyldisiloxane see Section 2.2.1. For potassium fluoride and tetraethylammonium bromide see Section 2.2.1.

Phenoxyacetic anhydride. To a stirred solution of phenoxyacetic acid (30.4 g, Janssen) in dry ether [150 ml, distilled from phosphorus pentoxide (30 g/l)] and dry dioxan (150 ml) add dicyclohexylcarbodiimide (21 g, Baker). Continue stirring for 3 h at room temperature and for 1 h at 0°C. Remove the precipitated dicyclohexylurea by filtration. Concentrate the filtrate to a colourless oil and crystallise from dry ether (200 ml) to afford pure phenoxyacetic anhydride. Yield 27 g, m.p. 67 – 69°C.

Diphenylacetic anhydride. To a stirred solution of diphenylacetic acid (42.2 g, Merck) in dry ether (150 ml) and dry dioxan (150 ml) add dicyclohexylcarbodiimide (21 g, Baker). Continue stirring for 3 h at room temperature and for 1 h at 0°C. Remove the precipitated dicyclohexylurea by filtration. Concentrate the filtrate to a colourless oil and crystallise from n-pentane (500 ml) to produce pure diphenylacetic anhydride. Yield 37 g, m.p. 96°C.

For melting points, t.l.c., column chromatography and n.m.r. spectroscopy see Section 2.1.1.

2.4.2 *2'-O-Phenoxyacetyl-3',5'-O-(tetraisopropyldisiloxane-1,3-diyl)guanosine*

(i) To a stirred suspension of dry guanosine (14.15 g; 50 mmol) in dry dimethylformamide (250 ml) add dropwise over a period of 15 min a solution of 1,3-dichloro-1,1,3,3-tetraisopropyldisiloxane (20 ml; 70 mmol) in dry dimethylformamide (50 ml) and dry pyridine (50 ml).

(ii) After 1 h at 50°C add an extra portion of 1,3-dichloro-1,1,3,3-tetraisopropyldisiloxane (1 ml; 5 mmol) and stir the clear solution for another 30 min at 50°C. T.l.c. (system B) should indicate complete conversion of the starting material.

(iii) Allow the reaction mixture to cool to room temperature and add phenoxyacetic anhydride (13) (157 g; 55 mmol). After stirring for 10 min at room temperature t.l.c. (system B) should indicate that the second step has gone to completion.
(iv) Pour the reaction mixture into a mixture of sodium bicarbonate (25 g), ice (500 g) and methylene chloride (500 ml).
(v) Separate the organic layer and extract the aqueous layer with methylene chloride (2 x 250 ml).
(vi) Wash the combined organic layers with water (500 ml), dry over magnesium sulphate (50 g), concentrate under reduced pressure to give a colourless oil, and crystallise from methanol (500 ml) to afford pure 2'-O-phenoxyacetyl-3',5'-O-(tetraisopropyldisiloxane-1,3-diyl)guanosine. Yield 28.0 g (85%), m.p. 199–200°C, R_f 0.35 (system A), ^{1}H-n.m.r. (CDCl$_3$): 12.7 (N-H, broad s), 7.87 (H$_8$, s), 7.5–7.0 (phenoxyacetyl, 5H, m), 6.6 (NH$_2$, broad s), 5.98 (H$_1$', s), 4.81 (phenoxyacetyl, 2H, s), ^{13}C-n.m.r. (CDCl$_3$): 158.8 (C$_6$), 153.9 (C$_2$), 150.7 (C$_4$), 135.6 (C$_8$), 117.5 (C$_5$), 86.7 (C$_1$'), 81.8 (C$_4$'), 68.9 (C$_3$'), 65.2 (C$_2$'), 60.3 (C$_5$'), 167.5, 157.8. 129.5, 121.8, 114.8, 50.3 (phenoxyacetyl), 17.3, 16.9, 13.4, 12.9, 12.6 (TIPS).

2.4.3 3',5'-O-(Tetraisopropyldisiloxane-1,3-diyl)-2-N-diphenylacetylguanosine (14)

(i) Stir a solution of 2'-O-phenyoxyacetyl-3',5'-O-(tetraisopropyldisiloxane-1,3-diyl)guanosine (16.5 g; 25 mmol) and diphenylacetic anhydride (34.5 g; 75 mmol) in dry pyridine (250 ml) at 50°C for 2 h. T.l.c. (system A) should show complete conversion of the starting material.
(ii) After addition of water (10 ml) concentrate the reaction mixture to a small volume (50 ml) under reduced pressure, dissolve in methylene chloride (500 ml), wash with an aqueous solution of sodium bicarbonate (1 M; 250 ml) and water (250 ml), dry over magnesium sulphate (25 g) and concentrate to a brown oil.
(iii) Dissolve the oil in methanolic ammonia (250 ml). After 20 min at room temperature t.l.c. (system A) should show complete removal of the phenoxyacetyl group.
(iv) Remove the methanolic ammonia by evaporation and apply the residue to a column of silica gel (100 g; 7 x 6.5 cm) packed in methylene chloride.
(v) Elute the column with methylene chloride/methanol (500 ml, 99.5:0.5; 500 ml, 99:1).
(vi) Collect the appropriate fractions and concentrate to a small volume (50 ml).
(vii) Crystallise the crude product from methanol/water (250 ml; 4:1) to afford pure 3',5'-O-(tetraisopropyldisiloxane-1,3-diyl)-2-N-diphenylacetylguanosine. Yield 13.7 g (75%), m.p. 184–187°C, R_f 0.57 (system A), ^{1}H-n.m.r. (CDCl$_3$/CD$_3$OD): 8.02 (H$_8$, s), 7.3 (diphenylacetyl, 10H, m), 5.80 (H$_1$', s), 5.12 (diphenylacetyl, 1H, s), ^{13}C-n.m.r. (CDCl$_3$/CD$_3$OD): 156.1

(C_6), 148.2 (C_2), 148.0 (C_4), 136.9 (C_8), 121.3 (C_5), 89.4 (C_1'), 82.0 (C_4'), 75.6 (C_3'), 69.3 (C_2'), 60.5 (C_5'), 174.3, 138.1, 128.0–127.8, 58.4 (diphenylacetyl), 17.4, 17.0, 13.7, 13.2, 13.1, 12.7 (TIPS).

2.4.4 2'-O-Tetrahydropyranyl-2-N-diphenylacetylguanosine (1d)

(i) To a stirred solution of 3',5'-O-(tetraisopropyldisiloxane-1,3-diyl)-2-N-diphenylacetylguanosine (7.2 g; 10 mmol) in dry dioxan (100 ml) add dry 2,3-dihydropyran (10 ml) and mesitylenesulphonic acid (0.2 g; 1 mmol).

(ii) After stirring for 2 h at room temperature add another portion of 2,3-dihydropyran (5 ml). After 4 h, t.l.c. (system A) should indicate complete conversion of the starting material.

(iii) Neutralise the reaction mixture with methanolic ammonia and concentrate under reduced pressure to a small volume (20 ml).

(iv) Dilute with methylene chloride (200 ml) and wash with an aqueous solution of sodium bicarbonate (1 M; 100 ml) and water (100 ml).

(v) Dry the organic layer over magnesium sulphate (10 g) and concentrate to give an oil.

(vi) Dissolve the oil, which contains both diastereoisomers of 2'-O-tetrahydropyranyl-3',5'-O-(tetraisopropyldisiloxane-1,3-diyl)-2-N-diphenylacetylguanosine, in acetonitrile (100 ml).

(vii) Add an aqueous solution of potassium fluoride (5 M; 10 ml) and tetraethylammonium bromide (10 g) and stir the resulting mixture for 1 h at 60°C. T.l.c. (system A) should show complete removal of the silyl group.

(viii) Concentrate the reaction mixture under reduced pressure to a small volume (20 ml).

(ix) Take up the residue in methylene chloride (200 ml) and wash with an aqueous solution of sodium bicarbonate (1 M; 100 ml) and water (100 ml).

(x) Dry the organic layer with magnesium sulphate (10 g), concentrate to a small volume (20 ml) and triturate with petroleum-ether (40–60°C, 200 ml).

(xi) Filter off the precipitate, dissolve in methylene chloride (10 ml) and apply to a column of silica gel (150 g; 12 x 3.5 cm) packed in methylene chloride.

(xii) Elute the column with methylene chloride/methanol (300 ml, 98:2; 300 ml, 96:4, 500 ml, 9:1) and collect the fractions containing both diastereoisomers and re-chromatograph on a column of silica gel (150 g; 12 x 3.5 cm) packed in methylene chloride.

(xiii) Elute the column with methylene chloride/methanol (300 ml, 98:2; 300 ml, 96:4; 500 ml, 9:1).

(xiv) Collect the fractions containing one of two diastereoisomers in its pure form and concentrate to a glass.

Yield of the high-running diastereoisomer of 2'-O-tetrahydropyranyl-2-N-diphenylacetylguanosine is 1.4 g (30%), R_f 0.34 (system A), ^{1}H-n.m.r. (CDCl$_3$): 12.21 (N-H, broad s), 11.87 (N-H, broad s), 7.81 (H_8, s), 7.3–7.1 (diphenylacetyl, 10H, m), 5.79 (H_1', d, J 6.0 Hz), 5.45 (diphenylacetyl, 1H, s), ^{13}C-n.m.r. (CDCl$_3$): 155.4 (C_6), 147.8 (C_2), 147.6 (C_4), 137.7 (C_8), 121.9 (C_5), 88.0 (C_1'),

81.2 (C$_4$'), 71.3 (C$_3$'), 65.3 (C$_2$'), 62.3 (C$_5$'), 174.7, 139.3, 128.7−127.6, 58.2 (diphenylacetyl), 101.0, 86.1, 30.7, 24.8, 19.7 (THP). Yield of the low-running diastereoisomer is 1.2 g (20%), R_f 0.25 (system A), ^{1}H-n.m.r. (CDCl$_3$): 12.25 (N-H, broad s), 11.40 (N-H, broad s), 7.95 (H$_8$, s), 7.4−7.1 (diphenylacetyl, 10H, m), 5.94 (H$_1$', d, J 3.0 Hz), 5.58 (diphenylacetyl, 1H, s), ^{13}C-n.m.r. (CDCl$_3$): 155.6 (C$_6$), 148.0 (C$_2$), 147.6 (C$_4$), 137.7 (C$_8$), 121.9 (C$_5$), 88.4 (C$_1$'), 78.9 (C$_4$'), 69.9 (C$_3$'), 67.0 (C$_2$'), 62.8 (C$_5$'), 175.0, 139.6, 128.6−127.5, 57.8 (diphenylacetyl), 99.5, 85.6, 30.0, 23.9, 19.6 (THP).

2.5 Synthesis of 2'-O-Tetrahydropyranyl-2-N-acetyl-6-O-(4-nitrophenylethyl)-guanosine (1e)

2.5.1 Reagents and Solvents

Dry guanosine (Waldhof, FRG) before use for 2 h at 50°C over phosphorus pentoxide *in vacuo*.

For dioxan, pyridine and 2,3-dihydropyran see Section 2.1.1. For dimethylformamide and acetonitrile see Section 2.2.1. For methanolic ammonia see Section 2.1.1. For triethylammonium bicarbonate buffer (2 M) see Section 2.2.1. For mesitylenesulphonic acid see Section 2.2.1. For 1,3-dichloro-1,1,3,3-tetraisopropyldisiloxane see Section 2.2.1. For potassium fluoride and tetraethylammonium bromide see Section 2.2.1. 4-Nitrophenylethanol (Aldrich), diethyl azadicarboxylate (Aldrich), triphenylphosphine (Fluka) and acetic anhydride (analytical grade, Baker) are used without further purification.

For melting points, t.l.c. and column chromatography see Section 2.1.1. ^{1}H-n.m.r. spectra are measured at 200 MHz using a JEOL JNM-FX 200 spectrometer. ^{13}C-n.m.r. spectra are measured at 50.3 MHz using a JEOL JNM-FX 200 spectrometer. Chemical shifts are given in p.p.m. (δ) relative to tetramethylsilane (TMS) as internal standard.

2.5.2 *2',3',5'-Tri-O-acetyl-2-N-acetylguanosine* (15)

(i) Heat a mixture of dry guanosine (14.15 g; 50 mmol), dry pyridine (250 ml) and acetic anhydride (17.5 ml) under gentle reflux for 2 h.

(ii) Add another portion of acetic anhydride (7.5 ml) and continue heating for another 1.5 h.

(iii) Add an extra portion of acetic anhydride (5 ml) and heat the reaction mixture for 1 h.

(iv) Allow the reaction mixture to cool down to room temperature and, after the addition of methanol (100 ml), concentrate under reduced pressure to give an oil.

(v) Take up the oil in methylene chloride (1000 ml) and wash with an aqueous solution of sodium bicarbonate (1 M; 500 ml) and water (500 ml).

(vi) Dry the organic layer over magnesium sulphate (50 g), concentrate to a small volume (100 ml) and triturate with petroleum-ether (40−60°C, 1000 ml).

(vii) Remove the petroleum-ether by decantation and concentrate the residue to give a glass.
(viii) Dissolve in methylene chloride (100 ml) and apply to a column of silica gel (300 g; 12 x 7 cm) packed in methylene chloride.
(ix) Elute the column with methylene chloride/methanol (2000 ml, 97:3).
(x) Collect the appropriate fractions and concentrate to give a yellow glass. Yield of pure 2',3',5'-tri-O-acetyl-2-N-acetylguanosine 17.7 g (76%), R_f 0.34 (system A), ^{1}H-n.m.r. (CDCl$_3$): 12.1 (N-H, broad s), 10.0 (N-H, broad s), 7.84 (H$_8$, s), 6.0 (H$_1$', d, J 3.0 Hz), 2.32, 2.14, 2.06, 2.02 (4 x - acetyl, 12H, 4 x s).

2.5.3 *2-N-Acetyl-6-O-(4-nitrophenylethyl)guanosine* (7)

(i) Co-evaporate a solution of 2',3',5'-tri-O-acetyl-2-N-acetylguanosine (22.6 g; 50 mmol), 4-nitrophenylethanol (12.6 g; 75 mmol) and triphenylphosphine (19.6 g; 75 mmol) in dry dioxan (400 ml) with dioxan (2 x 200 ml).
(ii) Add diethyl azadicarboxylate (11.6 ml; 75 mmol) and stir the resulting solution at room temperature for 30 min under the exclusion of mixture. T.l.c. (system A) should indicate complete conversion of the starting material.
(iii) Take up the reaction mixture in methylene chloride (1000 ml) and wash with triethylammonium bicarbonate buffer (1 M; 500 ml) and water (500 ml).
(iv) Dry the organic layer over magnesium sulphate (50 g) and concentrate to a brown oil.
(v) Dissolve the oil in methanol/triethylamine/water (1000 ml, 3:3:1) and stir at room temperature for 5 h. T.l.c. (system A) should show complete removal of the three O-acetyl groups.
(vi) Remove the volatile compounds by evaporation.
(vii) Apply the residue to a column of silica gel (300 g; 12 x 7 cm) packed in methylene chloride.
(viii) Elute the column with methylene chloride/methanol (500 ml, 98:2; 500 ml, 95:5, 1000 ml, 9:1; 1000 ml, 8:2).
(ix) Collect all fractions containing product and re-chromatograph on a column of silica gel (300 g; 12 x 7 cm) packed in methylene chloride.
(x) Elute the column with methylene chloride/methanol (500 ml, 98:2; 500 ml, 95:5; 1000 ml, 9:1; 1000 ml, 8:2).
(xi) Collect the appropriate fractions and concentrate to a yellow oil.
(xii) Crystallise from ethanol (1000 ml). Yield 18.2 g (77%), m.p. 182 – 183°C, R_f 0.14 (system A), ^{1}H-n.m.r. (C$_2$D$_6$SO): 10.47 (N-H, broad s), 8.45 (H$_8$, s), 8.19 (4-nitrophenylethyl, 2H, d, J 8.0 Hz), 7.65 (4-nitrophenylethyl, 2H, d, J 8.0 Hz), 5.89 (H$_1$', d, J 5.9 Hz), 5.49, 5.20 (2'-OH,3'-OH, 2 x d, J 5.0 Hz), 4.98 (5'-OH, t, J 5.8 Hz), 4.78 (4-nitrophenylethyl, 2H, t, J 6.9 Hz), 3.33 (4-nitrophenylethyl, 2H, t, J 6.9 Hz), 2.20 (acetyl, 3H, s).

Chemical Synthesis of Small Oligoribonucleotides

2.5.4 3',5',O-(Tetraisopropyldisiloxane-1,3-diyl)-2-N-acetyl-6-O-(4-nitrophenylethyl)guanosine

(i) To a stirred solution of 2-N-acetyl-6-O-(4-nitrophenylethyl)guanosine (23.75 g; 50 mmol) in dry pyridine (250 ml) add dropwise over a period of 15 min at room temperature 1,3-dichloro-1,1,3,3-tetraisopropyldisiloxane (18 ml; 62.5 mmol).

(ii) After stirring for another 15 min at room temperature t.l.c. (system A) should indicate complete conversion of the starting material.

(iii) Neutralise the reaction mixture with triethylammonium bicarbonate buffer (2 M; 75 ml), concentrate under reduced pressure to a small volume (100 ml), dissolve in methylene chloride (1000 ml) and wash with an aqueous solution of sodium bicarbonate (1 M; 500 ml) and water (500 ml).

(iv) Dry the organic layer over magnesium sulphate (50 g) and concentrate to a colourless oil.

(v) Co-evaporate with toluene (3 x 100 ml) to give a white glass.

Yield of crude 3',5'-O-(tetraisopropyldisiloxane-1,3-diyl)-2-N-acetyl-6-O-(4-nitrophenylethyl)guanosine, which can be used without further purification in the next step, is 34.0 g (95%), R_f 0.58 (system A), ^{1}H-n.m.r. (CDCl$_3$): 8.63 (N-H, broad s), 8.17 (4-nitrophenylethyl, 2H, d, J 8.0 Hz), 8.04 (H_8, s), 7.96 (N-H, s), 7.50 (4-nitrophenylethyl, 2H, d, J 8.0 Hz), 6.00 ($H_1{'}$, s), 4.77 (4-nitrophenylethyl, 2H, t, J 7.0 Hz), 3.32 (4-nitrophenylethyl, 2H, t, J 7.0 Hz), 2.56 (acetyl, 3H, s).

2.5.5 2'-O-Tetrahydropyranyl-2-N-acetyl-6-O-(4-nitrophenylethyl)guanosine (1e)

(i) To a stirred solution of crude 3',5',O-(tetraisopropyldisiloxane-1,3-diyl)-2-N-acetyl-6-O-(4-nitrophenylethyl)guanosine (34.0 g; 48 mmol), obtained as described above, in dry dioxan (250 ml) add dry 2,3-dihydropyran (25 ml) and mesitylenesulphonic acid (1.0 g). Keep the reaction mixture at room temperature for 2 h. T.l.c. (system A) should indicate complete conversion of the starting material.

(ii) Neutralise the reaction mixture with methanolic ammonia and concentrate under reduced pressure to a small volume (100 ml).

(iii) Dissolve in methylene chloride (1000 ml) and wash with an aqueous solution of sodium bicarbonate (1 M; 500 ml) and water (500 ml).

(iv) Dry the organic layer with magnesium sulphate (50 g) and concentrate.

(v) Dissolve the resulting oil, which contains both diastereoisomers of 2'-O-tetrahydropyranyl-3',5'-O-(tetraisopropyldisiloxane-1,3-diyl)-2-N-acetyl-6-O-(4-nitrophenylethyl)guanosine, in acetonitrile (500 ml), and add an aqueous solution of potassium fluoride (5 M; 50 ml) and tetraethylammonium bromide (50 g).

(vi) Stir the resulting mixture for 1 h at 60°C. T.l.c. (system A) should show complete removal of the silyl group.

(vii) Concentrate the reaction mixture under reduced pressure to a small volume.

(viii) Take up the residue in methylene chloride (1000 ml) and wash with an aqueous solution of sodium bicarbonate (1 M; 500 ml) and water (500 ml).
(ix) Dry the organic layer with magnesium sulphate (50 g), concentrate to a small volume (50 ml) and triturate with petroleum-ether (40–60°C, 1000 ml).
(x) Filter off the precipitate, dissolve in methylene chloride (50 ml) and apply to a column of silica gel (300 g; 12 x 7 cm) packed in methylene chloride.
(xi) Elute the column with methylene chloride (400 ml) and methylene chloride/ methanol (400 ml, 99:1; 400 ml, 98:2; 400 ml, 96:4; 400 ml, 9:1).
(xii) Collect the fractions containing both diastereoisomers and re-chromatograph on a column of silica gel (300 g; 12 x 7 cm) packed in methylene chloride.
(xiii) Elute the column with methylene chloride (400 ml) and methylene chloride/methanol (400 ml, 99:1; 400 ml, 98:2; 400 ml, 96:4; 400 ml, 9:1).
(xiv) Collect the fractions containing one of the two isomers in its pure form and concentrate to a glass.

Yield of the high-running diastereoisomer of 2'-O-tetrahydropyranyl-2-N-acetyl-6-O-(4-nitrophenylethyl)guanosine is 8.5 g (30%), R_f 0.46 (system A), ^{1}H-n.m.r. (CDCl$_3$): 8.16 (4-nitrophenylethyl, 2H, d, J 8.6 Hz), 8.4 (N-H, s), 7.95 (H$_8$, s), 7.51 (4-nitrophenylethyl, 2H, d, J 8.6 Hz), 5.92 (H$_1$', d, J 7.3 Hz), 4.80 (4-nitrophenylethyl, 2H, t, J 7.0 Hz), 3.32 (4-nitrophenylethyl, 2H, t, J 7.0 Hz), 2.44 (acetyl, 3H, s). ^{13}C-n.m.r. (CDCl$_3$): 160.7 (C$_6$), 152.1 (C$_2$), 151.7 (C$_4$), 142.0 (C$_8$), 119.3 (C$_5$), 88.7 (C$_1$'), 81.7 (C$_4$'), 71.7 (C$_3$'), 65.8 (C$_2$'), 62.9 (C$_5$'), 146.9, 145.6, 130.0, 123.7, 67.8, 35.3 (4-nitrophenylethyl), 169.9, 24.8 (acetyl), 102.0, 86.7, 30.7, 25.1, 21.2 (THP). Yield of the low-running diastereoisomer is 10.0 g (36%), R_f 0.35 (system A), ^{1}H-n.m.r. (CDCl$_3$): 8.4 (N-H, s), 8.17 (4-nitrophenylethyl, 2H, d, J 8.4 Hz), 8.01 (H$_8$, s), 7.53 (4-nitrophenylethyl, 2H, d, J 8.4 Hz), 6.02 (H$_1$', d, J 6.2 Hz), 4.81 (4-nitrophenylethyl, 2H, t, J 6.8 Hz), 3.33 (4-nitrophenylethyl, 2H, t, J 8.4 Hz), 2.40 (acetyl, 3H, s), ^{13}C-n.m.r. (CDCl$_3$): 161.0 (C$_6$), 153.0 (C$_2$), 152.7 (C$_4$), 142.7 (C$_8$), 118.7 (C$_5$), 88.8 (C$_1$'), 79.5 (C$_4$'), 70.3 (C$_3$'), 63.1 (C$_2$'), 62.4 (C$_5$'), 147.4, 130.7, 124.1, 67.5, 35.4 (4-nitrophenylethyl), 171.7, 24.8 (acetyl), 99.9, 86.7, 30.7, 25.6, 19.8 (THP).

3. SYNTHESIS OF 2'-O-TETRAHYDROPYRANYL-5'-O-(2-HEXADECYLOXY-4',4"-DIMETHOXYTRITYL)RIBONUCLEOSIDE MONOMERS

3.1 Reagents and Solvents

Use methanol (analytical grade, Baker) without further purification. Dry ether and tetrahydrofuran (Baker) by heating under reflux with phosphorus pentoxide (30 g/l) for 2 h and distil. For pyridine see Section 2.1.1. For dimethylformamide see Section 2.2.1. Dry cyclohexane (Baker) with sodium (Merck). Magnesium curls (Merck) and hexadecyl chloride (Fluka) are used without further purification. Distil 4-bromoanisole (Janssen), methyl salicylate (Janssen) and acetyl chloride (Baker) before use.

Carry out t.l.c. on silica gel (Schleicher and Schüll, TLC-Ready Plastic Sheets

F1500 LS 254). Use the following solvent systems: A, methylene chloride/methanol (92:8); C, ethyl acetate/methanol (98:2); for n.m.r. spectroscopy see Section 2.5.1.

3.2 Synthesis of 2-Hexadecyloxy-4′,4″-dimethoxytrityl Alcohol

3.2.1 *2-(Hexadecyloxy)benzoic Acid Methyl Ester* (16)

(i) Dissolve sodium (4.6 g; 0.2 mol) in methanol (200 ml, analytical grade).
(ii) Concentrate the resulting clear solution to dryness and re-dissolve in dry dimethylformamide (400 ml).
(iii) Add methyl salicylate (26.0 ml; 0.2 mol) and homogenise the solution by swirling.
(iv) Now add hexadecyl chloride (70.0 ml; 0.23 mol) and heat the reaction mixture under reflux for 1 h.
(v) Remove the solvent by evaporation under reduced pressure, take up the residue in ether (1000 ml) and wash with water (500 ml).
(vi) Separate the organic layer, dry over magnesium sulphate (50 g) and concentrate to a colourless oil.
(vii) Crystallise from ether/ethanol/methanol (500 ml, 1:1:1). The yield of pure 2-(hexadecyloxy)benzoic acid methyl ester is 60 g (80%), m.p. 46−49°C.

3.2.2 *2-Hexadecyloxy-4′,4″-dimethoxytrityl Alcohol*

(i) Add a solution of 4-bromoanisole (50 g; 0.27 mmol) in dry ether (2000 ml) dropwise to a stirred mixture of magnesium curls (8 g; 0.33 mol) and dry ether (200 ml).
(ii) Add a solution of 2-(hexadecyloxy)benzoic acid methyl ester (3.76 g; 0.1 mol) in dry tetrahydrofuran (200 ml) dropwise to the stirred reaction mixture at 0°C.
(iii) Heat the mixture under reflux for 1 h and then pour into an aqueous solution of ammonium chloride (2 M; 500 ml).
(iv) Separate the organic layer and extract the aqueous layer with ether (2 x 500 ml).
(v) Dry the combined organic layers over magnesium sulphate (50 g) and concentrate to a colourless oil.
(vi) Crystallise the residue from ethanol (500 ml) to afford pure 2-hexadecyloxy-4′,4″-dimethoxytrityl alcohol.

Yield 40 g (74%), m.p. 46−49°C, R_f 0.77 (system A), R_f 0.68 (system C). After spraying with concentrated sulphuric acid/methanol (1:4), the trityl alcohol shows up as a bright red spot. ^{1}H-n.m.r. (CDCl$_3$): 7.15−6.35 (m, 4H, [4xH(Ph-OC$_{16}$H$_{33}$)], 7.05, 6.70 (2 x d, J 9.0 Hz, 4H, [4xH(Ph-OCH$_3$)], 5.2 [s, 1H,(C-OH)], 3.77 [t, J 6.0 Hz, 2H, (PhO-CH$_2$-C$_{15}$H$_{31}$)], 3.68 [s, 6H, (PhO-CH$_3$)], 1.5−1.0 [m, 28H, (OCH$_2$-C$_{14}$H$_{28}$-CH$_3$)], 0.87 [t, J 4.5 Hz, 3H, (C$_{15}$H$_{30}$-CH$_3$)], ^{13}C-n.m.r. (CDCl$_3$): 158.4 (Ph-)C-OCH$_3$), 156.7 [(Ph-)C-OC$_{16}$H$_{33}$], 139.2 [(CH$_3$OPh-)C-C], 135.7 [(C$_{16}$H$_{33}$OPh-)C-C], 130.0 [(C$_{16}$H$_{33}$OPh-)C-H], 128.8 [(CH$_3$OPh-)C-H], 128.7 [(C$_{16}$H$_{33}$OPh-)C-H], 120.1 [(C$_{16}$H$_{33}$OPh-)C-H], 112.8 [(CH$_3$OPh-)C-H], 112.3 [(C$_{16}$H$_{33}$OPh-)C-H], 81.4 (Ph$_3$-C), 68.3 (PhO-CH$_2$-C$_{15}$H$_{31}$), 55.1 (PhO-CH$_3$), 31.9, 29.7−29.0, 25.7, 22.7 (OCH$_2$-C$_{14}$H$_{28}$-CH$_3$), 14.1 (C$_{15}$H$_{30}$-CH$_3$).

3.3 Synthesis of the Ribonucleoside Monomers

3.3.1 2'-O-Tetrahydropyranyl-5'-O-(2-hexadecyloxy-4',4"-dimethoxytrityl)uridine (2a)

(i) Heat a clear solution of 2-hexadecyloxy-4',4"-dimethoxytrityl alcohol (3.93 g; 7.0 mmol) and acetyl chloride (5.0 ml; 70 mmol) in dry cyclohexane (40 ml) under reflux for 1 h.

(ii) Remove the solvent by evaporation under reduced pressure and co-evaporate the remaining residue, in order to remove the last traces of acetyl chloride, with dry cyclohexane (3 x 20 ml) and finally once with dry pyridine (20 ml).

(iii) Dissolve the residue in dry pyridine (10 ml) and add it to a solution of the high-running diastereoisomer of 2'-O-tetrahydropyranyluridine (1.64 g; 5.0 mmol) in dry pyridine (20 ml).

(iv) Stir the red coloured solution for 2 h at room temperature. T.l.c. (systems A and C) should indicate the disappearance of the 2'-O-tetrahydropyranyl-uridine.

(v) Add a drop of water and concentrate the reaction mixture under reduced pressure to a small volume (10 ml).

(vi) Take up in methylene chloride (100 ml) and wash with an aqueous solution of sodium bicarbonate (1 M; 50 ml) and water (50 ml).

(vii) Dry the organic layer over magnesium sulphate (5 g) and concentrate to give a yellow oil.

(viii) Co-evaporate with toluene (2 x 25 ml) and dissolve the residue in a mixture of ethyl acetate, petroleum-ether (40–60°C) and triethylamine (5 ml, 200:200:1) and apply to a column of silica gel (100 g; 7 x 6.5 cm) packed in the same mixture.

(ix) Elute the column with ethyl acetate/petroleum-ether (40–60°C)/triethylamine (500 ml, 200:200:1; 500 ml, 200:100:1; 1000 ml, 400:100:1).

(x) Collect the appropriate fractions (t.l.c. system C) and evaporate to produce homogeneous 2'-O-tetrahydropyranyl-5'-O-(2-hexadecyloxy-4',4"-dimethoxytrityl)-uridine as a white glass.

Yield 37.0 g (85%), R_f 0.68 (system A), 0.53 (system C), ^{1}H-n.m.r. (CDCl$_3$/pyridine-d$_5$): 10.60 (N-H, broad, s), 7.90 (H$_6$, d, J 8.0 Hz), 5.95 (H$_1$', d, J 2.6 Hz), 5.32 (H$_5$, d, J 8.0 Hz), ^{13}C-n.m.r. (CDCl$_3$/pyridine-d$_5$): 163.9 (C$_4$), 150.7 (C$_2$), 140.1 (C$_6$), 102.1 (C$_5$), 88.3 (C$_1$'), 80.9 (C$_4$'), 69.0 (C$_3$'), 64.5 (C$_2$'), 61.5 (C$_5$'). According to the above described procedure the other 5'-O-(2-hexadecyloxy-4',4"-dimethoxytrityl)-protected monoribonucleosides can be obtained in good yields.

3.3.2 2'-O-Tetrahydropyranyl-5'-O-(2-hexadecyloxy-4',4"-dimethoxytrityl)-4-N-anisoylcytidine (2b)

Yield 4.50 g (90%), R_f 0.74 (system A), 0.32 (system C), ^{1}H-n.m.r. (CDCl$_3$/pyridine-d$_5$): 8.98 (N-H, broad s), 8.49 (H$_6$, d, J 7.7 Hz), 7.78 (H$_5$, d, J 7.7 Hz), 5.96 (H$_1$', s), ^{13}C-n.m.r. (CDCl$_3$/pyridine-d$_5$): 164.2 (C$_4$), 155.8 (C$_2$), 145.6 (C$_6$), 101.0 (C$_5$), 91.0 (C$_1$'), 84.1 (C$_4$'), 68.3 (C$_3$'), 65.5 (C$_2$'), 61.8 (C$_5$').

3.3.3 2'-O-Tetrahydropyranyl-5'-O-(2-hexadecyloxy-4',4"-dimethoxytrityl)-6-N-benzoyladenosine (2c)

Yield 4.10 g (84%), R_f 0.75 (system A), 0.50 (system C). ^{1}H-n.m.r. (CDCl$_3$/pyridine-d$_5$): 9.48 (N-H, broad s), 8.75 (H$_8$, s), 8.23 (H$_2$, s), 6.22 (H$_1'$, d, J 5.0 Hz), ^{13}C-n.m.r. (CDCl$_3$/pyridine-d$_5$): 152.6 (C$_6$), 152.6 (C$_2$), 151.8 (C$_4$), 141.9 (C$_8$), 123.8 (C$_5$), 87.2 (C$_1'$), 80.6 (C$_4'$), 71.0 (C$_3'$), 67.5 (C$_2'$), 53.1 (C$_5'$).

3.3.4 2'-O-Tetrahydropyranyl-5'-O-(2-hexadecyloxy-4',4"-dimethoxytrityl)-2-N-diphenylacetylguanosine (2d)

Yield 4.38 g (79%), R_f 0.73 (system A), 0.48 (system C), ^{1}H-n.m.r. (CDCl$_3$/pyridine-$_5$): 12.25 (N-H, broad s), 11.9 (N-H), broad s), 7.84 (H$_8$, s), 5.89 (H$_1'$, d, J 6.6 Hz), ^{13}C-n.m.r. (CDCl$_3$/pyridine-d$_5$): 155.7 (C$_6$), 152.9 (C$_2$), 152.3 (C$_4$), 140.7 (C$_8$), 123.7 (C$_5$), 86.8 (C$_1'$), 80.5 (C$_4'$), 71.1 (C$_3'$), 65.5 (C$_2'$), 63.4 (C$_5'$).

3.3.5 2'-O-Tetrahydropyranyl-5'-O-(2-O-hexadecyloxy-4',4"-dimethoxytrityl)-2-N-acetyl-6-O-(4-nitrophenylethyl)guanosine (2e)

Yield 4.90 g (90%), R_f 0.77 (system A), 0.50 (system C), ^{1}H-n.m.r. (CDCl$_3$/pyridine-d$_5$): 8.3 (N-H, s), 8.0 (H$_8$, s), 6.08 (H$_1'$, d, J 5.6 Hz), ^{13}C-n.m.r. (CDCl$_3$/pyridine-d$_5$): 160.6 (C$_6$), 152.9 (C$_2$), 152.3 (C$_4$), 140.7 (C$_8$), 123.7 (C$_5$), 86.8 (C$_1'$), 80.5 (C$_4'$), 71.1 (C$_3'$), 67.5 (C$_2'$), 63.4 (C$_5'$).

4. SYNTHESIS OF A FULLY PROTECTED SHORT OLIGORIBONUCLEOTIDE

4.1 Reagents and Solvents

Distil 2-chlorophenol (Janssen) and phosphoryl chloride (Merck) before use. Dry N-methylimidazole (Janssen) by heating under reflux with calcium hydride (5 g/l) and distil under diminished pressure.

Anhydrous pyridine. Heat dry pyridine (see Section 2.1.1) under reflux with toluene-*p*-sulphonyl chloride (50 g/l) for 1 h and distil.

Anhydrous dioxan. Heat dry dioxan (see Section 2.1.1) under reflux with lithium aluminium hydride (5 g/l) for 1 h and distil. For triethylammonium bicarbonate buffer (2 M) see Section 2.2.1. For 1-hydroxybenzotriazole see Section 2.2.1. For t.l.c. see Section 2.1.1. Carry out column chromatography on silica gel (Merck, Kieselgel 7729) or on Sephadex LH20 (Pharmacia) (17). ^{31}P-n.m.r. spectra are measured at 40.48 MHz using a JEOL JNMPFT 100 spectrometer. Chemical shifts are given in p.p.m. (δ) relative to 85% H$_3$PO$_4$ as external standard.

4.2 Synthesis of the Phosphorylating Agent: 2-Chlorophenyl-O,O-bis[1-benzotriazolyl]phosphate (3)

4.2.1 *2-Chlorophenyl Phosphorodichloridate (18)*

(i) Heat a solution of 2-chlorophenol (258 g; 2 mol) and phosphoryl chloride (900 ml; 10 mol) and dry N-methylimidazole (1 ml) under reflux for 16 h.

(ii) Remove excess phosphoryl chloride by evaporation and obtain 2-chlorophenyl phosphorodichloridate by distillation under diminished pressure.
(iii) Re-distillation (b.p. 160°C, 20 mm Hg) produces pure 2-chlorophenyl phosphorodichloridate. Yield 370 g (75%), ^{31}P-n.m.r. (CDCl$_3$) 3.62 p.p.m.

4.2.2 *2-Chlorophenyl-O,O-bis[1-benzotriazolyl]phosphate* (3) (8)

(i) Add a solution of 2-chlorophenyl phosphorodichloridate (2.45 g; 10.0 mmol) in anhydrous dioxan (8 ml) dropwise to a stirred mixture of dry 1-hydroxybenzotriazole (2.70 g; 20.0 mmol), anhydrous pyridine (1.6 ml; 20 mmol) and anhydrous dioxan (40 ml) at room temperature.
(ii) After the addition stir the reaction mixture for 1 h at room temperature and remove the precipitated pyridinium hydrochloric acid salt by filtration under anhydrous conditions to give a solution of 2-chlorophenyl-O,O-bis[1-benzotriazolyl]phosphate in dioxan (0.2 M), ^{31}P-n.m.r. (CD$_3$CN): −8.4 p.p.m.

4.3. Synthesis of Dimer (5) (9)

(i) Co-evaporate 2′-O-tetrahydropyranyl-5′-O-(2-hexadecyloxy-4′,4″-dimethoxytrityl)uridine (high-running diastereoisomer, 2.57 g; 3.0 mmol) with anhydrous pyridine (3 x 10 ml) and add a solution of freshly prepared 2-chlorophenyl-O,O-bis[1-benzotriazolyl]phosphate in dioxan (0.2 M; 16.5 ml) to the viscous residue.
(ii) Stir the reaction mixture for 30 min at room temperature under the exclusion of moisture. T.l.c. (system A) should indicate complete conversion of the starting material.
(iii) Co-evaporate 2′-O-(tetrahydropyranyl)-6-N-benzoyladenosine (low-running diastereoisomer, 1.77 g; 3.9 mmol) with anhydrous pyridine (3 x 10 ml) and transfer to the reaction mixture under anhydrous conditions.
(iv) After 5 min add N-methylimidazole (1.2 ml; 15 mmol) and continue stirring for 30 min at room temperature. T.l.c. (system A) should indicate the disappearance of trityl-positive baseline material.
(v) After the addition of triethylammonium bicarbonate buffer (1 M; 0.2 ml), dilute the reaction mixture with methylene chloride (200 ml) and wash the solution twice with triethylammonium bicarbonate buffer (1 M; 100 ml, 0.1 M; 100 ml).
(vi) Dry the organic layer over magnesium sulphate (10 g), concentrate to a small volume (10 ml) and finally triturate with petroleum-ether (40−60°C, 200 ml).
(vii) Filter off the precipitate and dissolve in methylene chloride/triethylamine (10 ml, 400:1) and apply to a column of silica gel (50 g; 7 x 3.5 cm) packed in the same solvent mixture.
(viii) Elute the column with methylene chloride/methanol/triethylamine (200 ml, 400:4:1; 200 ml, 400:8:1; 200 ml, 400:12:1; 200 ml, 400:16:1; 200 ml, 400:40:1).
(ix) Collect the fractions containing pure dimer 5 and concentrate to a colourless glass. Yield 3.30 g; (2.20 mmol, 73%), R_f 0.55 (system A).

4.4 Synthesis of Trimer (6)

(i) Treat dimer 5 (3.30 g; 2.20 mmol) as described in Section 4.3 with 2-chlorophenyl-O,O-bis[1-benzotriazolyl]phosphate (0.2 M; 12.1 ml), 2'-O-tetrahydropyranyl-2-N-acetyl-6-O-(4-nitrophenylethyl)guanosine (low-running diastereoisomer, 1.60 g; 2.86 mmol) and N-methylimidazole (0.9 ml; 11 mmol).

(ii) Carry out work-up and purify by silica gel column chromatography (50 g; 7 x 3.5 cm) as described in Section 4.3 to produce homogeneous trimer 6 as a colourless glass. Yield 3.64 g (1.63 mmol, 74%), R_f 0.53 (system A).

4.5 Synthesis of Tetramer (7)

(i) Treat trimer 6 (3.64 g; 1.63 mmol) as described in Section 4.3 with 2-chlorophenyl-O,O-bis[1-benzotriazolyl]phosphate (0.2 M; 8.9 ml), 2'-O-tetrahydropyranyl-2-N-acetyl-6-O-(4-nitrophenylethyl)guanosine (1.18 g; 2.12 mmol) and N-methylimidazole (0.5 ml; 8.2 mmol).

(ii) After the addition of N-methylimidazole stir the reaction mixture for 1 h at room temperature.

(iii) Carry out work-up and purify by silica gel column chromatography (50 g; 7 x 3.5 cm) as described in Section 4.3, to produce homogeneous tetramer 7 as a colourless glass. Yield 3.67 g (1.24 mmol, 76%), R_f 0.52) (system A).

4.6. Synthesis of Pentamer (8)

(i) Treat tetramer 7 (3.67 g; 1.24 mmol) as described in Section 4.3 with 2-chlorophenyl-0,0-bis[1-benzotriazolyl]phosphate (0.2 M; 6.8 ml), 2'-O-tetrahydropyranyl-6-N-benzoyladenosine (0.734 g; 1.61 mmol) and N-methylimidazole (0.5 ml; 6.2 mmol).

(ii) After the addition of N-methylimidazole stir the reaction mixture for 1.5 h at room temperature.

(iii) Carry out work-up and purify by silica gel column chromatography (50 g, 7 x 3.5 cm) as described in Section 4.3 to produce pentamer 8. The product may be partially contaminated with 2'-O-tetrahydropyranyl-6-N-benzoyladenosine.

(iv) Collect the contaminated fraction, concentrate to a glass and dissolve in methylene chloride/methanol/triethylamine (5 ml, 200:100:1) and apply to a column of Sephadex LH 20 (3 x 100 cm) packed in the same solvent mixture.

(v) Elute the column with this mixture also.

(vi) Complete separation of pentamer from monomer should be realised within 1 h. Collect the appropriate fractions and concentrate to give homogeneous pentamer 8 as a colourless glass. Yield 3.02 g (0.84 mmol, 68%), R_f 0.52 (system A).

4.7 Synthesis of Hexamer (9a)

(i) Treat pentamer 8 (3.02 g; 0.84 mmol) as described in Section 4.3 with 2-chlorophenyl-O,O-bis[1-benzotriazolyl]phosphate (0.2 M; 4.6 ml),

2'-O-tetrahydropyranyluridine (0.358 g; 1.09 mmol) and N-methylimidazole (0.35 ml; 4.2 mmol).
(ii) After the addition of N-methylimidazole stir the reaction mixture for 2 h at room temperature.
(iii) Work-up and purification on silica gel column chromatography (50 g; 7 x 3.5 cm) is described in Section 4.3 and produces homogeneous hexamer **9a** as a white glass. Yield 2.57 g (0.63 mmol, 75%), R_f 0.50 (system A).

5. DEBLOCKING OF THE FULLY PROTECTED HEXARIBONUCLEOTIDE 9a

5.1 Reagents and Solvents

Toluene-*p*-sulphonic acid monohydrate (Baker), acetic anhydride (analytical grade, Baker), acetic acid (analytical grade, Baker), concentrated ammonia (14.8 M, analytical grade, Baker), *syn*-pyridine-2-carboxaldoxime (Janssen), N^1,N^1,N^3,N^3-tetramethylguanidine (Eastman Kodak) and 1,8-diazabicyclo[5.4.0]undecene (Janssen) are used without purification. For dioxan, pyridine and acetonitrile see Section 2.2.1. Use sterile water and glassware during the whole deblocking process.

Cation-exchange resin (sodium-form). Pass a solution of sodium hydroxide (2 N; 100 ml) over a column of cation-exchange resin (Dowex 50 W x 8, 100 – 200 mesh; Fluka, 44514, H$^+$-form; 1.5 x 5 cm) and wash the column with sterile water until the pH is 7.0. Carry out column chromatography on Sephadex G50 (Pharmacia).

H.p.l.c. analysis. Use a weak anion-exchange resin, Permaphase AAX (Du Pont), dry packed into a stainless steel column (2.1 mm x 1 m) (see Chapter 5 and Appendix 2). Elute at 50°C using a linear gradient, starting with buffer A (0.005 M KH$_2$PO$_4$, pH 4.1) and applying 3% buffer B (0.1 M KH$_2$PO$_4$, 1.0 M KCl, pH 4.5) per min. Flow-rate is 1 ml/min. Pressure is 80 kp/cm^2.

5.2 Removal of the 2-Hexadecyloxy-4',4''-dimethoxytrityl Group from Hexamer 9a

(i) Dissolve the fully protected hexamer **9a** (409 mg; 0.1 mmol) in a stirred solution of toluene-*p*-sulphonic acid monohydrate (10 g) in methylene chloride/ methanol (200 ml, 7:3) at 0°C (19). Immediately the solution should become intensely red coloured.
(ii) After 15 min at 0°C pour the reaction mixture into a stirred aqueous solution of sodium bicarbonate (1 M; 400 ml). The red colour disappears within a few seconds and when the evolution of carbon dioxide gas ceases, separate the organic layer and wash with water (400 ml).
(iii) Dry over magnesium sulphate (25 g), concentrate to a small volume (10 ml), and triturate with petroleum-ether (40 – 60°C, 500 ml).
(iv) Filter off the precipitate, wash with petroleum-ether (100 ml), dissolve in methylene chloride (500 ml) and concentrate to a colourless glass. Yield is

344 mg (97%). T.l.c. of the product R_f 0.38; system A) should reveal the absence of baseline-material, and the 2-O-hexadecyloxy-4',4''-dimethoxytrityl alcohol.

5.3 Acetylation of the Free Hydroxyl Groups of Hexamer 9b

(i) Dissolve the glass obtained under 5.2 (hexamer **9b**) (344 mg) in a mixture of dry pyridine (10 ml), acetic anhydride (1 ml) and N-methylimidazole (0.2 ml).
(ii) Stir the solution for 2 h at room temperature.
(iii) Add a drop of triethylammonium bicarbonate buffer (1 M).
(iv) Concentrate the mixture to a small volume (3 ml), dissolve in methylene chloride (100 ml) and wash with an aqueous solution of sodium bicarbonate (1 M; 50 ml) and water (50 ml).
(v) Separate the organic layer, dry over magnesium sulphate (5 g) and concentrate to a colourless oil.
(vi) Triturate the oil with petroleum-ether (40–60°C, 200 ml).
(vii) Filter off the precipitate dissolve in methylene chloride (50 ml) and concentrate to give a colourless glass. Yield 336 mg (96%), R_f 0.40 (system A).

5.4 Removal of the Base-labile Groups from Hexamer 9c

(i) Dissolve the glass obtained under 5.3 (hexamer 9c) in a solution of *syn*-pyridine-2-carboxaldoxime (20) (702 mg; 5.75 mmol) and N^1,N^1,N^3,N^3-tetramethylguanidine (575 mg; 5.0 mmol) in dry acetonitrile (7.5 ml) (21) and dry dioxan (7.5 ml).
(ii) Stir the reaction mixture for 24 h at room temperature under the exclusion of moisture.
(iii) Take up the reaction mixture in concentrated ammonia (14.8 M; 200 ml), leave for 48 h at 50°C, concentrate to a small volume (2 ml) and co-evaporate with dry pyridine (3 x 20 ml).
(iv) Add a solution of 1,8-diazabicyclo[5.4.0]undecene (152 mg; 1.0 mmol) in dry pyridine (2 ml) to the residue and stir the solution under the exclusion of moisture for 16 h at room temperature.
(v) Bring the reaction mixture to pH 6 by the addition of an aqueous solution of acetic acid (5 M).
(vi) Add a few drops of concentrated ammonia (14.8 M) and concentrate the reaction mixture to a small volume, then co-evaporate with water (3 x 20 ml). The residue (10 ml) contains undissolvable material which should be removed by filtration.
(vii) Wash the filtrate with ether (3 x 20 ml) to give a clear aqueous solution of the crude partially deblocked hexamer **9d**.

5.5 Purification of the Partially Deblocked Hexamer 9d

(i) Apply the aqueous solution of the partially deblocked hexamer **9d** obtained in Section 5.4 to a column of Sephadex G50 (2 x 200 cm) packed in triethylammonium bicarbonate buffer (0.5 M).
(ii) Elute the column with the same buffer at a flow rate of 14 ml/h.

(iii) Collect fractions of 3 ml and pool those containing pure product (as monitored by h.p.l.c.), and concentrate to a small volume (5 ml) to produce an aqueous solution of hexamer still containing the acid-labile 2'-O-tetrahydropyranyl groups.

5.6 Removal of the 2'-O-Tetrahydropyranyl Groups from Hexamer 9d

(i) Acidify the solution of the hexamer **9d** obtained in Section 5.5 with hydrochloric acid (0.01 N; 100 ml) and adjust the pH of the clear solution to 2.00 by the addition of hydrochloric acid (0.1 N).
(ii) Leave the mixture for 16 h at 20°C, and neutralise with aqueous ammonia (5 M) until the pH has become 8.00.
(iii) Apply the product, after concentration to a small volume (5 ml), to a column of cation-exchange resin (sodium-form; 1.5 x 5 cm) packed in sterile water (pH 7.0).
(iv) Carry out elution of the column with sterile water (pH 7.0).
(v) Collect u.v.-positive (254 nm) material, concentrate to a small volume (5 ml) and lyophilise to give the homogeneous fully deblocked hexamer **10** as a fluffy solid.

A detailed description of methods developed for the enzymatic analysis of synthetic RNA fragments by h.p.l.c. and other techniques is presented in Chapter 5.

6. THE PREPARATION OF LONGER RNA FRAGMENTS

6.1. Synthesis of a Hexadecamer RNA Fragment

The methodology described in the previous sections produces partially protected (3'-OH-free) RNA fragments, which can be used as building blocks for the assembly of longer RNA fragments. This is exemplified by the synthesis of the hexadecamer AUCCUAUUUUUAGGAU (*Figure 3*). For this purpose three fragments [i.e., the 5'-terminal hexamer I, the non-terminal tetramer II and the 3'-terminal hexamer III (**9a**)] are synthesised by the methods described above. The introduction of the two phosphotriester linkages between the fragments I, II and III is then carried out as follows.

Tetramer II is phosphorylated with the bifunctional phosphorylating agent **3** to afford intermediate II' which, in turn, is hydrolysed to give the 3'-phosphorylated tetramer II" (R_8 = 2-chlorophenyl; $R^2 = C_{16}$-DMTr). The latter is condensed, in the presence of the activating agent 2,4,6-trimethylbenzenesulphonyl-3-nitro-1,2,4-triazolide (MSNT; see Chapter 4), with the 3'-terminal hexamer III" to afford the fully protected decamer IV. The hexamer III" is obtained by acetylation of III, to give III' (R_7 = acetyl), followed by the removal of the acid-labile R_2-group with *p*-toluenesulphonic acid. Deprotection of the acid-labile group (R_2) from the decamer IV affords IV' which is coupled, in the presence of MSNT, with the 5'-terminal hexamer I", obtained in the same way as described for the preparation of II", to give the fully protected hexadecamer V. Complete deblocking of V followed by purification (see Section 5) gives the hexadecamer as a homogeneous fluffy solid. This block-condensation method is

Chemical Synthesis of Small Oligoribonucleotides

ApUpCpCpUpApUpApUpUpUpUpApGpGpApU

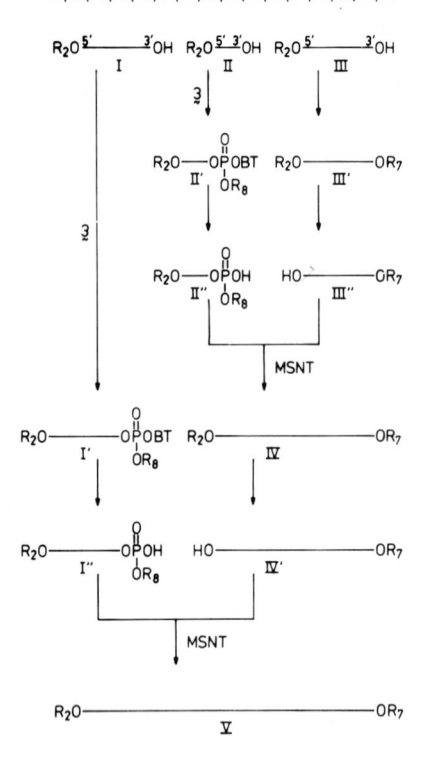

[Figure 3 scheme showing block condensation synthesis with intermediates I, II, III, I', II', III', I'', II'', III'', IV, IV', V using MSNT coupling]

Figure 3. Synthesis of a hexadecamer RNA fragment *via* block condensation.

especially attractive for the synthesis of RNA fragments in relatively large quantities (100 – 200 mg) which are necessary when carrying out extended bio-physical studies.

6.1.2. Synthesis of the Fully Protected Decaribonucleotide IV

(i) Treat the partially protected ribotetranucleotide II (sequence UUUU), carrying a 5'-O-[2-hexadecyloxy-4',4"-dimethoxytrityl] group (0.4 mmol) with phosphorylating agent **3** (0.2 M; 2.8 mmol) as described in Section 4.3.
(ii) Stir the reaction mixture for 30 min at room temperature with exclusion of moisture. T.l.c. (system A) should indicate complete conversion of the starting material.
(iii) Add triethylammonium bicarbonate buffer (1 M; 2 ml).
(iv) Dilute the reaction mixture with methylene chloride (100 ml) and wash the solution with triethylammonium bicarbonate buffer (1 M; 50 ml, 0.1 M; 50 ml).
(v) Concentrate the organic layer to a small volume and co-evaporate once with anhydrous pyridine (10 ml).

(vi) Add the partially protected hexaribonucleotide III″ (sequence UAGGAU; 0.3 mmol) obtained after acetylation of fragment III (**9a**) to Yield III′, followed by the removal of the 2-hexadecyloxy-4,4″-dimethoxytrityl group (see Section 6.1.3), and co-evaporate the mixture twice with anhydrous pyridine (2 x 10 ml).

(vii) Add MSNT (0.56 mmol) and stir the reaction mixture for 30 min at room temperature with exclusion of moisture.

(viii) Add a second portion of MSNT (0.08 mmol) and stir the reaction mixture for another 30 min. T.l.c. (systems A and B) should indicate complete disappearance of the trityl-positive baseline-material.

(ix) Work-up, as described in Section 4.3, produces the fully protected decaribonucleotide IV as a glass.

6.1.3 *Removal of the 2-Hexadecyloxy-4,4″-dimethoxytrityl Group from IV*

(i) Add the crude reaction product IV which is dissolved in cold (0°C) methylene chloride/methanol (20 ml, 7:3) to a stirred solution of toluene-*p*-sulphonic acid monohydrate (10 g) in methylene chloride/methanol (200 ml, 7:3) at 0°C. Immediately the solution should become intensely red coloured.

(ii) After 15 min at 0°C, pour the reaction mixture into a stirred aqueous solution of sodium bicarbonate (1 M, 400 ml). The red colour disappears within a few seconds.

(iii) When the evolution of carbon dioxide gas ceases, separate the organic layer and wash with water (400 ml).

(iv) Dry over magnesium sulphate (25 g), concentrate to a small volume (10 ml) and triturate with petroleum-ether (40 – 60°C, 500 ml).

(v) Filter off the precipitate, wash with petroleum-ether (40 – 60°C, 100 ml) and dissolve in methylene chloride/methanol/triethylamine (5 ml, 200:100:1).

(vi) Apply the residue to a column of Sephadex LH60 (3 x 100 cm) packed in the same solvent mixture.

(vii) Elute the column with the same solvent mixture. Collect the appropriate fractions and concentrate to give the partially protected decamer IV′ (R_f 0.38, system A) which should be free of the starting compounds I and II″. Yield 1.03 g (0.22 mmol, 72%, starting from II″).

6.1.4 *Synthesis of the Fully Protected Hexaribonucleotide V*

(i) Treat the partially protected hexaribonucleotide I (sequence AUCCUA), carrying a 5′-O-[2-hexadecyloxy-4′,4″-dimethoxytrityl] group (0.4 mmol) with phosphorylating agent **3** (0.2 M; 2.8 mmol) as described in Section 4.3.

(ii) Stir the reaction mixture for 30 min at room temperature with exclusion of moisture. T.l.c. (system A) should indicate conversion of the starting material.

(iii) Add triethylammonium bicarbonate buffer (1 M; 2 ml).

(iv) Dilute the reaction mixture with methylene chloride (100 ml) and wash the solution with triethylammonium bicarbonate buffer (1 M; 50 ml, 0.1 M; 50 ml).
(v) Concentrate the organic layer to a small volume and co-evaporate once with anhydrous pyridine (10 ml).
(vi) Add the partially protected decaribonucleotide IV' (sequence UUUUUAGGAU, 0.2 mmol) carrying a free 5'-OH group and co-evaporate the mixture twice with anhydrous pyridine (2 x 10 ml).
(vii) Add MSNT (0.56 mmol) and stir the reaction mixture for 1 h at room temperature with exclusion of moisture.
(viii) Add a second portion of MSNT (0.16 mmol) and stir the reaction mixture for another 2 h. T.l.c. (systems A and B) should indicate complete disappearance of the trityl-positive baseline-material.
(ix) Work-up as described in Section 4.3 produces the fully protected hexadecaribonucleotide V as a glass.
(x) Dissolve the product in methylene chloride/methanol/triethylamine (5 ml, 200:100:1) and apply it on to a column of Sephadex LH60 (3 x 100 cm) packed in the same solvent mixture.
(xi) Elute the column with the same mixture and collect the appropriate fractions, and concentrate to give the fully protected hexadecaribonucleotide V (R_f 0.5, system A). Yield 1.43 g (0.115 mmol, 77%).

6.2 Future Trends

Preliminary experiments indicate that the introduction of the two phosphotriester linkages between the three fragments I, II and III can also be carried out directly starting from the 3'-phosphotriester intermediates I' and II". Further, the feasibility of a step-wise introduction of phosphotriester linkages using agent 3 between an immobilised ribonucleoside and a properly protected ribonucleoside also seems to be very promising. This solid-phase approach will be a very economical route to the synthesis of RNA fragments on a small scale.

7. ACKNOWLEDGEMENTS

This work was supported by the Netherlands Foundation for Chemical Research (SON), with financial aid from the Netherlands Organisation for the Advancement of Pure Research (ZWO). We wish to thank Mr. R.P. van de Woestijne and Mr. A. Fidder for their technical assistance and Mr. F. Lefeber for recording the n.m.r. spectra.

8. REFERENCES

1. Reese,C.B. (1978) *Tetrahedron*, **34**, 3143.
2. Ohtsuka,E., Ikehara,M. and Söll,D. (1982) *Nucleic Acids Res.*, **10**, 6553.
3. Görtz,H.H. and Seliger,H. (1981) *Angew. Chem. (Engl. Ed.)*, **20**, 681.
4. Griffin,B.E., Jarman,M. and Reese,C.B. (1968) *Tetrahedron*, **24**, 639.
5. Fromageot,H.P.M., Griffin,B.E., Reese,C.B. and Sulston,J.E. (1967) *Tetrahedron*, **23**, 2315.
6. Markiewicz,W.T. (1979) *J. Chem. Res.*, 24.
7. Himmelsbach,F., Schulz,B.S., Trichtinger,T., Charubala,R. and Pfleiderer,W. (1984) *Tetrahedron*, **40**, 59.

8. Van der Marel,G., van Boeckel,C.A.A., Wille,G. and van Boom,J.H. (1981) *Tetrahedron Lett.*, **22**, 3887.
9. Wreesmann,C.T.J., Fidder,A., van der Marel,G.A. and van Boom,J.H. (1983) *Nucleic Acids Res.*, **11**, 8399.
10. Fromageot,H.P.M., Griffin,M., Reese,C.B., Sulston,J.E. and Trentham,D.R. (1966) *Tetrahedron*, **22**, 705.
11. Steinfeld,A.S., Naider,F. and Becker,J.M. (1979) *J. Chem. Res.*, 129.
12. Takaku,H., Yoshida,M. and Nomoto,T., (1983) *J. Org. Chem.*, **48**, 1399.
13. Kamimura,T., Masegi,T. and Hata,T. (1982) *Chem. Lett.*, 965.
14. Verdegaal,C.H.M., Jansse,P.L., de Rooij,J.F.M., Veeneman,G. and van Boom,J.H. (1981) *Recl. Trav. Chim. Pays-Bas*, **100**, 200.
15. Reese,C.B. and Saffhill,R. (1972) *J. Chem. Soc. Perkin Trans. I*, 2937.
16. Parker,R.A., Kariya,T., Grisar,J.M. and Petrow,V. (1977) *J. Med. Chem.*, **20**, 781.
17. Biernat,J., Wolter,A. and Köster,H. (1983) *Tetrahedron Lett.*, **24**, 751.
18. Owen,S.R., Reese,C.B., Ranson,C.J., van Boom,J.H. and Herscheid,J.D.H. (1974) *Synthesis*, 704.
19. Takaku,H., Nomoto,T., Sakamoto,Y. and Hata,T. (1979) *Chem. Lett.*, 1225.
20. Reese,C.B. and Zard,L. (1981) *Nucleic Acids Res.*, **9**, 4611.
21. Van Boeckel,C.A.A. and van Boom,J.H. (1980) *Tetrahedron Lett.*, **21**, 3705.

CHAPTER 8

Enzymatic Synthesis of Oligoribonucleotides

DOROTHY BECKETT AND OLKE C. UHLENBECK

1. INTRODUCTION

The availability of synthetic RNA molecules of defined sequence has greatly aided the investigation of the structure and functions of RNA. While RNA molecules can be synthesised by entirely chemical or enzymatic methods, a combination of both is often the best. In this chapter, we focus on two enzymatic methods that have made the synthesis of olgioribonucleotides accessible to any laboratory familiar with simple biochemical reactions. Reactions involving primer-dependent polynucleotide phosphorylase are summarised as a means of preparing short oligomer blocks. The combined use of T4 RNA ligase and polynucleotide kinase allows the joining of blocks to make long RNA molecules.

If comparatively modest amounts of RNA fragments are needed, the enzymatic approach to oligoribonucleotide synthesis has several advantages. Firstly, enzyme reactions tend to be more familiar to the biochemist or molecular biologist who will often use the oligonucleotides. Organic synthesis methods tend to require a larger commitment of expertise and materials. Secondly, due to the absence of chemical blocking groups, the mild reaction and work-up conditions and the high specificity of enzyme reactions, the products of enzymatic synthesis often tend to be more homogeneous than those of organic synthesis. This is important since, unlike DNA fragments which are often purified by cloning, synthetic RNA fragments are used directly in applications. Thirdly, enzymatic methods are well suited for semi-synthetic protocols where fragments of natural RNA molecules are used to prepare the molecule of interest. This permits synthetic efforts to be focussed on a narrow region of an RNA molecule for structure-function studies. For example, several groups have developed procedures to replace nucleotides in the anticodon loop of tRNAs (1 – 3). Finally, the substrate specificity of RNA ligase allows the facile incorporation of a variety of base- and sugar-modified nucleotides into an RNA chain without the development of special methods for each modification (4). The availability of such subtly modified RNAs may yield information about which substituent groups in the RNA molecule are necessary for biological function (5).

There are, of course, several disadvantages with enzymatic RNA synthesis. The inherent specificity of the enzymes to different oligonucleotide sequences greatly complicates their use. The conditions required to obtain optimal yield of a given reaction tend to be different for every sequence, thereby necessitating tedious trial reactions. In some cases, reaction yields are so low that a successful synthesis cannot be achieved and an alternate pathway must be developed. Finally, since all enzyme reactions can reverse, the accumulation of undesired side products as a result of reverse reactions can be a problem. Although enzymatic synthesis procedures can often be used to prepare modest (1 – 100 μg) amounts of oligoribonucleotides up to 20 residues long for biochemical studies, the scaling up of synthesis to obtain the amounts suitable for biophysical studies has often proven difficult. The losses encountered in the multiple purification steps and the expense of preparing large amounts of enzymes is a clear limitation. These problems may best be circumvented by combining chemical and enzymatic procedures. Chemical synthesis can be used to prepare blocks of the four normal nucleotides and enzymes can be used to join the blocks and introduce modified nucleotides.

2. GENERAL STRATEGY

The synthesis of a long RNA fragment uses T4 RNA ligase to join shorter RNA blocks. It is generally most efficient to use a branched pathway, where smaller fragments are joined to make larger ones and the larger ones are joined to make the final sequence. Sequential additions of short fragments from one end of the chain would require large quantities of the starting oligomers and necessitate more difficult separations of products from starting materials. Since the precise design of a pathway for a given sequence depends greatly on the availability of the shorter blocks, their preparation will be discussed first.

2.1 Synthesis of Short Blocks

Polynucleotide phosphorylase from *Micrococcus luteus* or *Escherichia coli* can be used to prepare homopolyribonucleotides and mixed co-polymers from nucleoside diphosphates (6). Limited alkaline hydrolysis of the homopolymers and separation of the resultant oligomers provides the homo-oligomers $(Ap)_n$, $(Up)_n$, $(Cp)_n$ and $(Gp)_n$ although the latter series are difficult to separate. Adjustment of the hydrolysis time provides some control over the average length of oligomer obtained. Total digestion of the appropriate co-polymer with either ribonuclease A or ribonuclease T1 and separation of the oligomers provides the $(Ap)_nUp$, $(Ap)_nCp$, $(Ap)_nGp$, $(Cp)_nGp$ and $(Up)_nGp$ series [the $(Gp)_nUp$ and $(Gp)_nCp$ series are difficult to separate]. The relative amount of each member of the series can be controlled by the ratio of nucleotides in the co-polymer (6).

The digestion of *M. luteus* polynucleotide phosphorylase with trypsin renders it primer-dependent so that the polymerisation of nucleoside-5'-diphosphates only occurs onto the free 3'-hydroxyl group of an added primer (7). At high ionic strengths, the polymerisation reaction can be controlled such that a limited number of nucleotides can be added (8,9). While this reaction has been used for the synthesis of a variety of block co-polymers (10), it is complicated by the fact that at longer times the polymerisation reaction reverses, leading to scrambling of product sequences. This difficulty is avoided if dimers are used as primers in the following reaction:

$$XpY + ppN \rightleftarrows XpY(pN)_n + P_i$$

Since dimers are poor primers compared with larger RNAs, the reaction proceeds in two stages. First, the reaction reaches equilibrium by making relatively long chains on a small number of the available dimer molecules. The equilibrium constant is roughly unity. Subsequently, as a result of multiple reverse and forward steps, the diphosphates redistribute themselves on all the dimer molecules, resulting in an average chain length which depends upon the input ratio of dimer and diphosphate. The redistribution step is much slower than the initial equilibrium step, but results in yields of 10–25% of the dimer in a given product. One advantage of this reaction is that the yield for each value of n is about the same. Longer oligomers can therefore be made as easily as shorter ones.

In some cases, very high yields of trimers can be obtained by including a sequence-specific ribonuclease in the polynucleotide phosphorylase reaction (9). After the second diphosphate is added to the dimer, the nuclease will cleave the phosphodiester linkage to produce a trimer with a 3' phosphate, which is no longer a primer in the reaction. For example, in the synthesis of ApCpGp

$$\text{ApC} + \text{ppG} \xrightarrow[\text{T1}]{\text{PNPase}} \text{ACGp} + \text{G} + \text{P}_i$$

the yield of product is greater than 90%. This reaction can also be used to make longer oligomers in a similar way provided that the primer is not sensitive to the nuclease.

Although many oligomers can be made by the methods described above, some cannot. One problem is that polynucleotide phosphorylase does not use dipyrimidine dimers as primers very efficiently, prohibiting the efficient synthesis of UpU(pA)$_n$ and similar series. A second problem is that longer oligomers with a high proportion of G residues tend to aggregate, prohibiting XpY(pG)$_n$ reactions. In *Table 1* the enzymatic synthesis of the 64 trimers is classified into three groups.

(i) Group 1: homo-oligomers and products of nuclease-assisted reactions that are easy to prepare in large (10 μmol or more) amounts;
(ii) Group 2: products of polymerisation reactions which are not difficult to make, but lower yields make larger scale synthesis less feasible.
(iii) Group 3: difficult to obtain in even modest amounts.

There are 20 trimers in group 1, 32 in group 2 and 12 in group 3. Since trimers are the minimal length acceptor used in the RNA ligase reaction, the information in *Table 1* guides the strategy of synthesis.

A variety of other sources of short RNA blocks are available that will not be discussed in this chapter. Methods for preparing oligoribonucleotides by chemical synthesis are described in the previous chapter. In some cases total nuclease digestion of specific RNA molecules can provide useful fragments. The reversal of nuclease reactions is an efficient means of making certain short oligomers (11). Finally, a wide variety of nucleoside-3',5'-bisphosphates, which are active donors in the RNA ligase reaction, can be prepared by direct phosphorylation of the nucleoside using pyrophosphoryl chloride (4).

Enzymatic Synthesis

Table 1. Synthesis of Trinucleotides.

			Central nucleotide					
			U	C	A	G		
			1	2	2	2	U	
			2	2	2	2	C	
		U	3	3	2	2	A	
			1	1	1	2	G	
			2	2	2	2	U	
			2	1	2	2	C	
		C	3	3	2	2	A	
5'			1	1	1	3	G	3'
nucleotide			2	2	1	1	U	nucleotide
			2	2	1	1	C	
		A	2	2	1	2	A	
			1	1	1	3	G	
			2	2	1	1	U	
			2	2	1	1	C	
		G	2	2	2	3	A	
			3	3	3	3	G	

1, easy to make in large amounts; 2, can be made in moderate amounts; 3, difficult to make.

2.2 Joining of Blocks with RNA Ligase

Oligomers obtained by one of the above methods are joined using T4 RNA ligase. The enzymatic properties of this enzyme have been reviewed recently (12) and more practical descriptions of its use are also available (13,14). The reaction mechanism consists of three steps.

(1) $E + ATP \rightleftarrows E-pA + pp_i$
(2) $E-pA + pN_nX \rightleftarrows E[A5'pp5'N_nX]$
(3) $E[A5'pp5'nX] + M_m \rightleftarrows M_mpN_nX + AMP + E$

The first step is the adenylylation of the enzyme by ATP. In the second step the AMP residue is transferred to the 5'-phosphate of a donor molecule, pN_nX, to form a phosphoanhydride linkage. The third step involves the adenylylated donor reacting with an acceptor molecule with 3'- and 5'-hydroxyls, M_m, to give the product M_mpN_nX. The presence of a blocking group X prevents addition to the 3' terminus of the donor by other donor molecules and thereby ensures a unique product. Removal of X from the product gives an acceptor molecule that can be used in a subsequent RNA ligase reaction. Alternatively, phosphorylation of the 5' terminus of the product with polynucleotide kinase yields a donor molecule for another RNA ligase reaction. Thus, the synthesis can proceed in either direction by a branched pathway.

The most convenient blocking group X is simply a 3' (or 2') terminal phosphate. It can be easily removed with bacterial or calf intestine alkaline phosphatase. It is already available on oligomers prepared by nuclease digestions or it can be introduced by periodate oxidation of the 3'-ribose followed by β-elimination of the 3'-base resulting in an oligomer one nucleotide shorter than the original. However, if 3'-phosphates are used as blocking groups, it is important to use T4 polynucleotide kinase derived from the T4 *pseT 1* mutant which lacks the 3'-phosphatase activity of wild-type T4 polynucleotide kinase (15).

The only difficulty with using 3'-phosphates as blocking groups is that in ligase reactions involving certain combinations of donors and acceptors, 3'-phosphorylated termini are preferential sites of reversal of the reaction resulting in additional undesired products (16). The most common example of reversal is the removal of a 3',5'-bisphosphate from the 3' end of relatively long (six or greater) oligomer when it reacts with adenylylated enzyme:

$$E - pA + (Np)_n Mp \rightarrow (Np)_{n-1} N + A5'pp5'Mp + E$$

Both products of this reverse reaction can react with other oligomers in the reaction mixture by the forward RNA ligase reaction to give undesired products. Since reversal occurs preferentially at a 3'-terminal phosphate, a variety of other 3'-blocking groups can be used to avoid this problem. These include the acid-labile O (α-methoxyethyl) (17) and ethoxymethylidine (18) groups and the photolabile *o*-nitrobenzyl group (19). In addition it has recently been shown (20) that the oligoribonucleotides prepared by chemical synthesis, that are partially deblocked such that all 2'-hydroxyls contain tetrahydropyranyl groups, are excellent substrates for polynucleotide kinase and active donors (but not acceptors) in the RNA ligase reaction. This provides a potentially important way to combine chemical and enzymatic synthesis.

RNA ligase does not react with all donors and acceptors with the same efficiency. The shortest active acceptor is a trinucleoside diphosphate and the shortest donor is a mononucleoside-5',3'-bisphosphate. Increasing the length of the oligonucleotide does not substantially affect the rate of the reaction. However the sequence of the acceptor and, to a lesser extent, the donor are important determinants of how well a ligation reaction will work. While systematic studies of reaction yields with different combinations of donor and acceptors are not complete (14), a general picture of how well different oligomers react has emerged. Donors in which the 5'-terminal nucleotide is a pyrimidine are slightly more reactive than when it is a purine. More importantly, acceptors where either of the two 3'-terminal nucleotides is a uridine are much less active than other acceptors.

The general strategy of how to synthesise an RNA molecule by enzymatic methods therefore involves several considerations. Firstly, the sequence should be broken into blocks that can be prepared efficiently with a blocking group on the donor molecules. Secondly, the ligation sites should avoid certain sequences and should not be carried out in regions of potential secondary structure. Thirdly, the ligation pathway should be branched to make optimal use of the short blocks. *Figure 1* gives an example of such a pathway for the synthesis of a fragment of R17 RNA, which is the site of translational repression by R17 coat protein (21). A particularly convenient consequence of such a branched pathway is that variants of the original sequence can be constructed without re-synthesising the whole molecule (22). The availability of sequence variants is often useful in structure-function studies of biologically important RNA molecules.

Enzymatic Synthesis

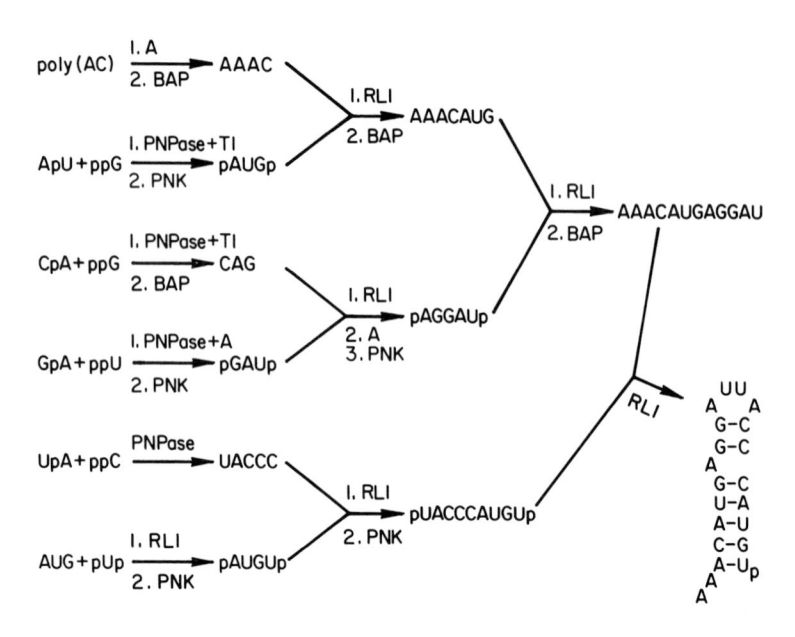

Figure 1. Branched synthetic scheme for the 21 nucleotide R17 coat protein-binding fragment (21). A: ribonuclease A, BAP: bacterial alkaline phosphatase, T1: ribonuclease T1, PNPase: polynucleotide phosphorylase, PNK: *pseT 1* polynucleotide kinase, RLI: RNA ligase.

3. HANDLING RNA FOR ENZYMATIC SYNTHESIS

3.1 Equipment

Although only minimal equipment is required to carry out enzymatic synthesis reactions (automatic pipettors, vortex mixer, and 14°C and 37°C water baths), a good deal of equipment is needed to analyse and purify the reaction products. The most important item is an h.p.l.c. system including programmable solvent delivery, a u.v. detector and an integrator. Other less important but useful equipment includes a tank for descending paper chromatography, a t.l.c. 'sandwich' and a gel electrophoresis apparatus. A good vacuum pump and condenser are needed for the rapid removal of solvents. This can be attached to a rotary evaporator (Büchi/Brinkmann Instruments) for reduction of volumes more than 10 ml, an Evapo-Mix (Buchler Instruments) for volumes between 10 and 0.2 ml and a Speed Vac Concentrator (Savant Instruments) for volumes less than 0.2 ml. The latter two are especially useful for reducing oligonucleotide solutions to dryness at the very bottom of a conical tube. A clinical centrifuge and, to a lesser extent, a microfuge (Brinkmann Instruments) are also useful for handling oligonucleotide solutions.

3.2 Buffers and Supplies

A major problem encountered in handling RNA fragments, that is not generally seen with DNA fragments, is degradation of the sample as a result of introduction of RNases through poor technique or contaminated reagents. Although the extreme vigilence required for RNA sequencing (23) can be relaxed when larger amounts of RNA are used, a number of precautions must be taken.

All glassware used for synthesis and work-ups must be washed with chromic acid, rinsed exhaustively with glass-distilled water, and baked for several hours at 300°C. Buffers should be prepared with glass-distilled water from reagents of highest purity, autoclaved and stored at −20°C. Disposable plastic pipette tips and tubes should be used. Autoclaving and siliconising them is not generally necessary unless very low RNA concentrations are used. Disposable gloves need to be used when a chance of contacting the RNA sample occurs.

Since ribonucleases are used in some enzymatic synthesis reactions, special care must be taken that they do not contaminate the rest of the laboratory. Weigh out and solubilise lyophilised nucleases in another laboratory. Carry out all nuclease reactions in disposable tubes. H.p.l.c. columns used to analyse nuclease reactions should be kept separate.

3.3 Separation and Purification of Oligonucleotides

A variety of separation methods is available to purify an oligonucleotide from a reaction mixture of an enzymatic synthesis. Of these, h.p.l.c. is undoubtedly the best for analysis of the course of a reaction and is also very useful for preparative purification. Several suitable systems are discussed in Chapter 5. Our laboratory prefers an ion-pair system using an octadecyl silica column and a solvent containing 5 mM tetrabutylammonium phosphate pH 7 (or acetate pH 5) in acetonitrile-water mixtures. Linear gradients in which the acetonitrile is varied from 20 to 40% result in oligomers eluting strictly according to charge and independent of composition. This results in a degree of predictability not available in other h.p.l.c. systems.

The presence of proteins in enzymatic oligonucleotide synthesis reactions can cause special problems in h.p.l.c. purification since denatured proteins will plug up h.p.l.c. lines. For an analytical injection of a few microlitres, centrifugation of the reaction mixture for several minutes in a microcentrifuge is sufficient. However, for preparative injections, it is important to deproteinise the reaction mixture since proteins will denature and precipitate in most h.p.l.c. solvents. A convenient way to deproteinise is to use small amounts of octadecyl silica resin which has been packed into a syringe barrel (C18 BOND-ELUTE, Analytichem International). These columns fit into a 12 ml conical test tube and each aliquot is forced through the column by centrifugation for 15 sec in a clinical centrifuge. Alternatively, the resin is packed in a cartridge which fits onto a syringe (Sep-Pak, Waters Associates, Inc.).

(i) Prepare the resin for use by successive 2 x 1 ml washes of water, 5 mM tetrabutylammonium phosphate solution (pH 7.0), 5 mM ammonium acetate solution, water, 90% acetonitrile, water and 5 mM ammonium acetate solution.

(ii) Make the reaction mixture 5 mM in tetrabutylammonium phosphate and apply to the column.

(iii) Wash the column successively with 2 x 1 ml of 5 mM ammonium acetate solution and water. Proteins, salts, nucleoside di- and triphosphates wash through the column, while oligomers stick. Oligomers shorter than 12 residues can be effectively eluted with two 1 ml washes of 90% acetonitrile. Since recoveries of oligomers longer than 12 are relatively low, deproteinisation of larger fragments is better done by phenol extraction and subsequent ethanol precipitation.

(iv) After the protein is removed from a preparative reaction, the reaction is resuspended in the starting h.p.l.c. buffer and injected.

Other oligonucleotide separation methods are often used in enzymatic synthesis. Descending chromatography on Whatman 3MM paper using mixtures of 1 M ammonium acetate solution and 95% ethanol are useful for moderate scale (μmol) preparation of oligomers less than six residues in length (21). Column chromatography on DEAE Sephadex A-25 using linear gradients of triethylammonium bicarbonate is useful for large-scale preparation of oligomers up to 10 residues. Much longer oligomers can be eluted with sodium chloride gradients. Polyacrylamide gel electrophoresis is useful to purify small amounts of radioactive oligomers with chain lengths of eight or more (24).

4. POLYNUCLEOTIDE PHOSPHORYLASE REACTIONS

4.1 Materials

Sources of materials and storage conditions are given in *Table 2*. The pH of solutions of dimers and nucleoside-5'-diphosphates must be adjusted to pH 7. Occasionally commercial preparations of nucleotides are contaminated with a ribonuclease. Purification by absorption to a small DEAE-Sephadex A-25 column and elution with triethylammonium bicarbonate is usually sufficient to solve this problem.

TMN, a 5-fold concentrated buffer for polynucleotide phosphorylase reactions contains 50 mM $MgCl_2$, 2.0 M NaCl and 1.0 M Tris-HCl (pH 8.2) at 37°C.

Primer-independent polynucleotide phosphorylase can be purified from *M. luteus* (25). It is converted to the primer-dependent form by a limited trypsin digestion (7). Both forms can be purchased from several sources (e.g., Pharmacia

Table 2. Materials for Polynucleotide Phosphorylase Reactions.

Material	Supplier	Concentration	Storage buffer
Dimers	S, P, C	25 mM (10 mM for GpG)	10 mM Tris pH 7.0
Nucleoside diphosphates	S, P, C	0.35 M	10 mM Tris pH 7.0
Ribonuclease A	S, B, W	1 mg/ml	10 mM Tris pH 7.0
Ribonuclease T1	C, W	1 mg/ml	10 mM Tris 1 mM $MgCl_2$ pH 7.0
E. coli alkaline phosphatase	W, C	1 mg/ml	10 mM Tris 10 mM $MgCl_2$ 10 mM NaCl pH 8.2

P, Pharmacia PL Biochemicals Inc.; B, Boehringer Mannheim GmbH Biochemica; S, Sigma Chemical Company; W, Worthington Biochemical Corporation; C, Calbiochem-Behring.

PL Biochemicals). A unit of activity corresponds to 1 μmol of ADP incorporated into polymer in 15 min (7). The enzyme is stored frozen in 10 mM Tris (pH 7.6), 1 mM EDTA, 100 mM NaCl at a concentration of at least 20 units/ml.

4.2 Equilibrium Polynucleotide Phosphorylase Reaction

(i) The reaction for the synthesis of $XpY(pN)_n$ contains: 5 mM of XpY; 35 mM of ppN; 20% (v/v) of TMN; 5 units/ml of polynucleotide phosphorylase.

(ii) Incubate at 37°C for 24 h if X and Y are purines, 36 – 48 h if one is a pyrimidine, and up to 72 h if both are pyrimidines. Although these conditions are generally successful for many combinations of dimers and diphosphates, it is prudent to carry out trial reactions to test the components and incubation time. These may be done as 25 μl reactions if the mixtures are sealed in capillary tubing to prevent evaporation. It is best to use a reaction expected to go well [such as $ApA(pC)_n$] to test out the components.

As mentioned previously, the reaction proceeds by a rapid polymerisation to form long polymers followed by a redistribution to give the oligomers. Thus, since a polymer is an intermediate in the reaction, the analytical system used to follow the progress should be able to detect the complete distribution of chain lengths or measure the absolute amount of given oligomer in an aliquot.

(iii) Adjust the enzyme concentration with the incubation time such that equilibrium is reached. Since the enzyme is active at 37°C for many days, the long incubations suggested above are in order to save enzyme. The use of more enzyme will result in reaching equilibrium more rapidly, but will not give higher yields.

The reaction conditions given in (i) yield a relatively broad chain length distribution with approximately equal molar amounts of trimer to heptamer (~10% of the dimer in each), and lesser amounts out to pentadecamer. As would be expected, changing the ratio of diphosphate to dimer will cause a corresponding change in the product length distribution. If trimer is the only oligomer desired, 12 mM ppN will give 20 – 30% of the dimer converted to trimer. As an extreme example, very efficient incorporation of radioactive diphosphate into trimer can be achieved when the diphosphate concentration is less than 0.1 mM (26). One advantage of using lower diphosphate concentrations is that less enzyme and shorter incubation times can generally be used since fewer polymerisation and phosphorolysis steps are required to reach equilibrium.

For some purification procedures, it is convenient to dephosphorylate the disphosphate with bacterial alkaline phosphatase after the reaction is complete. This can be done in the same reaction mixture in the following way.

(i) Inactivate the polynucleotide phosphorylase by heating to 90°C for 2 min. This prevents additional phosphorolysis when phosphate is released from the diphosphate during the phosphatase reaction.

Enzymatic Synthesis

(ii) Dilute the reaction mixture by adding two volume of water. Dilution is only required at diphosphate concentrations above 5 mM in order to prevent inhibition of alkaline phosphatase by the released phosphate.
(iii) Add bacterial alkaline phosphatase to a final concentration of 50 μg/ml.
(iv) Incubate for 3 – 4 h at 37°C.

4.3 Nuclease-assisted Reactions

(i) The reaction conditions for the synthesis of certain XpYpNp trimers contains: 5 mM of XpY; 35 mM of ppN; 20% (v/v) of TMN; 50 μg/ml of nuclease; 5 units/ml of polynucleotide phosphorylase.
(ii) Incubate for 24 h at 37°C. Generally, very high yields (>80%) can be expected. Higher enzyme concentrations and longer incubation times may have to be used if a dipyrimidine dimer is used. Since the nuclease is in substantial excess, no longer oligomers accumulate in the reaction. In RNase T1 reactions a trinucleotide with a 2′,3′ cyclic terminal phosphate is often seen early in the reaction, but is generally completely hydrolysed within 24 h.

5. POLYNUCLEOTIDE KINASE AND RNA LIGASE REACTIONS

5.1 Materials

Polynucleotide kinase can be purified from T4-infected *E. coli* (27). If a 3′-terminal phosphate is going to be used as a donor blocking group in RNA ligase reactions, it is important to use the *pseT 1* mutant strain of T4. This mutation inactivates the inherent 3′-specific phosphatase activity of wild-type polynucleotide kinase. Both wild-type and *pseT 1* polynucleotide kinase are available from several commercial sources (Pharmacia PL Biochemicals, New England Nuclear, Boehringer Mannheim).

A unit of polynucleotide kinase activity catalyses the transfer of 1 nmol of the γ-phosphate of ATP to the 5′-hydroxyl group of nuclease-treated DNA in 30 min at 37°C (27). Homogeneous enzyme is approximately 200 000 units/mg. Polynucleotide kinase is generally stored at about 2000 units/ml in 25 mM KH_2PO_4 (pH 7.0), 12.5 mM KCl, 2.5 mM dithiothreitol (DTT), 25 μM ATP and 50% (v/v) glycerol at −20°C. It is stable for several years under these conditions. At less than 1 μM ATP, the enzyme inactivates rapidly even at 4°C.

RNA ligase can be purified from the same batch of T4-infected *E. coli* used for polynucleotide kinase (28). The enzyme can also be purified from *E. coli* cells containing a recombinant plasmid (29). RNA ligase is available from several commercial sources (Pharmacia PL Biochemicals, Miles Biochemicals, New England Nuclear).

A unit of RNA ligase activity is defined as the incorporation of 1 nmol of [5′-^{32}P]pA$_n$ into a phosphomonoesterase-resistant form in 30 min at 37°C (28). Since this assay has not been found to be very reproducible, enzyme concentrations are often expressed in μg/ml. Homogeneous enzyme is approximately 2000 units/mg. The enzyme is stored at above 1 mg/ml in 20 mM Hepes (pH 7.5), 1 mM DTT, 10 mM $MgCl_2$ and 50% (v/v) glycerol at −20°C.

Use the same 5x buffer in both polynucleotide kinase and RNA ligase reactions. It contains 250 mM Hepes (pH 7.5) (at 14°C), 100 mM MgCl$_2$, 15 mM DTT, 50 μg/ml nuclease-free bovine serum albumin (Sigma Chemical Co., Pharmacia PL Biochemicals). Sterilise the buffer by filtration and store at −20°C. Store ATP in 10 mM Hepes (pH 7.5) as a 10 mM solution.

The various sources of oligomer blocks to be used in joining reactions have been described above. It is important that inhibitory contaminants are purified away from the oligomers before they are used in enzyme reactions. Ammonium, triethylammonium and tetrabutylammonium cations, phosphate anions and protein denaturants such as SDS or urea inhibit both enzymes. Small amounts (<5%) of organic solvents are not inhibitory. The C-18 purification method described in Section 3.3 is usually suitable for the removal of contaminants. Store oligomer solutions in water at 2 mM strand concentration or higher at −20°C.

5.2 Polynucleotide Kinase Reaction

(i) A suitable protocol for the quantitative 5′-phosphorylation of an oligonucleotide is: 1 mM of oligomer; 2 mM of ATP; 20% (v/v) of 5x buffer; 50−100 units/ml of T4 kinase.

(ii) Incubate at 37°C. The reaction is generally complete in 3 h. Since the enzyme slowly inactivates in the reaction, incubations longer than 6 h are usually not useful.

(iii) Terminate the reaction by heating to 90°C for 2 min. These relatively high oligomer concentrations make optimal use of the enzyme. Quantitative phosphorylation can also be achieved at oligomer concentrations as low as 10 μM. If this is done, the ATP concentration may be reduced proportionately, but the enzyme can only be reduced to 20 units/ml. The high oligomer concentrations also permit heat-inactivated kinase reactions to be added directly to an RNA ligase reaction (see below).

Since the rate of polynucleotide kinase reaction shows little dependence on the sequence of the oligomer, the above conditions should be suitable for most oligomers. Thus, if a phosphorylation is not complete, it generally means an inhibitor is present in the reaction. This problem can sometimes be overcome by diluting the reaction mixture with buffer and adding additional enzyme. Otherwise re-purification of the oligomer is necessary. In some cases, RNA secondary structure is known to inhibit polynucleotide kinase reactions (30). Due to the sensitivity of polynucleotide kinase to heat denaturation, the addition of RNA denaturants to alleviate this problem has not been successful.

5.3 RNA Ligase Reactions

The optimal reaction conditions for joining a pair of oligoribonucleotides with RNA ligase vary considerably depending upon the donor and acceptor pair. It is therefore important to carry out trial reactions in small volumes (10−20 μl) to optimise reaction conditions.

(i) The following conditions will give good yields for many donor and acceptor pairs and therefore are a good starting point: 0.5 mM of acceptor;

0.5 mM of donor; 1.0 mM of ATP; 20% (v/v) of 5x buffer; 100 µg/ml of RNA ligase. A heat-inactivated polynucleotide kinase reaction can be used as a source of phosphorylated donor although a purified donor is generally more active.

(ii) Incubate at 14°C. The reaction is usually complete in 6 h. Even if the reaction is not complete, little additional product forms at longer incubation times. If the yield is very high, it may be possible to reduce the enzyme concentration by as much as 10-fold for a preparative reaction.

If the above conditions do not give good yields, a number of things can be tried to improve the situation.

(1) The enzyme concentration can be increased. Since the relationship between enzyme concentration and reaction yield is not simple (31), as little as a 5-fold increase in the enzyme concentration can have a dramatic effect on the yield.

(2) The donor and acceptor concentrations can be reduced to as low as 0.1 mM. This will not only reduce the number of turnovers per enzyme molecule but will lower the concentration of possible enzyme inhibitors.

(3) The addition of 10–20% (v/v) of dimethylsulphoxide can substantially improve the yield in some cases (32).

(4) The incubation temperature can be varied between 0° and 37°C. Although 14°C is generally as good as any other temperature, some donor-acceptor pairs react more readily at other temperatures.

(5) Raising the pH of the reaction to pH 8.3 can sometimes improve the yield of product with respect to adenylylated donor (31). Other changes in the buffer conditions have not affected RNA joining reactions although manganese ion has been found to aid joining reactions with DNA acceptors (28).

(6) The ratio of ATP to donor concentrations is another important factor in the yield of product. An excess of ATP over donor is usually maintained to ensure that the equilibrium favours product formation. However, this excess is detrimental in reactions involving poor acceptors. In these cases, adenylylated donor formed in the second step of the mechanism dissociates from the enzyme before ligation to acceptor occurs. If the free enzyme becomes adenylylated by reaction with ATP before the adenylylated donor can re-bind, the system can become 'over-adenylylated' where all the enzyme and donor is adenylylated and no further reaction with acceptor can occur. This situation can be avoided by using ATP concentrations that are the same or a little less than the donor concentration. If this is done the reaction is often slower but higher yields can be obtained.

6. REFERENCES

1. Bruce,A.G. and Uhlenbeck,O.C. (1982) *Biochemistry (Wash.),* **21**, 855.
2. Vacher,J., Grosjean,H., de Henau,S., Finelli,J. and Buckingham,R.H. (1984) *Eur. J. Biochem.,* **138**, 77.
3. Shulman,L.H. and Pelka,H. (1983) *Nucleic Acids Res.,* **11**, 1439.

4. Barrio,J.R., Barrio,M.C.G., Leonard,N.J., England,T.E. and Uhlenbeck,O.C. (1978) *Biochemistry (Wash.)*, **17**, 2077.
5. Wittenberg,W.L. and Uhlenbeck,O.C. (1984) Submitted for publication.
6. Littauer,U.Z. and Soreq,H. (1982) in *The Enzymes,* Vol. **XV**, Academic Press, New York, p. 517.
7. Klee,C.B. (1969) *J. Biol. Chem.*, **244**, 2558.
8. Thach,R.E. and Doty,P. (1965) *Science (Wash.)*, **147**, 1310.
9. Lang,A. and Gassen,H.G. (1982) in *Chemical and Enzymatic Synthesis of Gene Fragments — A Laboratory Manual,* Gassen,H.G. and Lang,A. (eds.), Verlag Chemie, Weinheim, p. 149.
10. Martin,F.H., Uhlenbeck,O.C. and Doty,P. (1971) *J. Mol. Biol.*, **57**, 201.
11. Mohr,S.C. and Thach,R.E. (1969) *J. Biol. Chem.*, **244**, 6566.
12. Uhlenbeck,O.C. and Gumport,R.I. (1982) in *The Enzymes,* 3rd ed., Vol. **15**, Boyer,P.D. (ed.), Academic Press, New York, p. 31.
13. Romaniuk,P.J. and Uhlenbeck,O.C. (1983) *Methods Enzymol.*, **100**, 52.
14. Gumport,R.I. and Uhlenbeck,O.C. (1981) in *Gene Amplification and Analysis, Vol. 2 Analysis of Nucleic Acid Structure by Enzymatic Methods,* Chirikjian,J.G. and Papas,T.S. (eds.), Elsevier/North Holland, Amsterdam, p. 313.
15. Cameron,V., Soltis,D. and Uhlenbeck,O.C. (1978) *Nucleic Acids Res.*, **5**, 825.
16. Krug,M. and Uhlenbeck,O.C. (1982) *Biochemistry (Wash.)*, **21**, 1858.
17. Sninsky,J.J., Last,J.A. and Gilham,P.T. (1976) *Nucleic Acids Res.*, **3**, 3157.
18. Ohtsuka,E., Nishikawa,S., Markham,A.F., Tanaka,S., Miyake,T., Wakabayashi,T., Ikehara,M. and Sugiura,M. (1978) *Biochemistry (Wash.)*, **17**, 4984.
19. Ohtsuka,E., Vemura,H., Doi,T., Miyake,T., Nishikawa,S. and Ikehara,M. (1979) *Nucleic Acids Res.*, **8**, 601.
20. Romaniuk,P. Unpublished observations.
21. Krug,M., deHaseth,P.L. and Uhlenbeck,O.C. (1982) *Biochemistry (Wash.)*, **21**, 4713.
22. Carey,J., Lowary,P.T. and Uhlenbeck,O.C. (1983) *Biochemistry (Wash.)*, **22**, 4723.
23. D'Alessio,J.M. (1982) in *Gel Electrophoresis of Nucleic Acids,* Rickwood,D. and Hames,B.D. (eds.), IRL Press, Oxford and Washington, DC, p. 173.
24. Carey,J., Cameron,V., deHaseth,P.L. and Uhlenbeck,O.C. (1983) *Biochemistry (Wash.)*, **22**, 2601.
25. Klee,C.B. and Singer,M.F. (1968) *J. Biol. Chem.*, **243**, 923.
26. Uhlenbeck,O.C., Baller,J. and Doty,P. (1970) *Nature*, **225**, 508.
27. Soltis,D.A. and Uhlenbeck,O.C. (1982) *J. Biol. Chem.*, **257**, 11332.
28. Brennan,C.A., Manthey,A.E. and Gumport,R.I. (1983) *Methods Enzymol.*, **100**, 38.
29. Thøgersen,H.C., Morris,H.R., Rand,K.N. and Gait,M.J. (1985) *Eur. J. Biochem.*, **147**, 325.
30. Lillehaug,J.R., Kleppe,R.K. and Kleppe,K. (1976) *Biochemistry (Wash.)*, **15**, 1858.
31. Uhlenbeck,O.C. and Cameron,V. (1977) *Nucleic Acids Res.*, **4**, 85.
32. Bruce,A.G. and Uhlenbeck,O.C. (1978) *Nucleic Acids Res.*, **5**, 3665.

APPENDIX I

General Laboratory Techniques for Oligonucleotide Synthesis

BRIAN S. SPROAT and MICHAEL J. GAIT

1. SHORT COLUMN CHROMATOGRAPHY

A typical short column is depicted in *Figure 1*. The top of the column has a wide (preferably B34 or larger) Quickfit joint and two glass hooks and at the bottom of the column there is a flat glass sinter to retain the adsorbent. The column should be designed such that there is as little dead space as possible below the glass sinter. The column outlet consists of a male Luer fitting. As far as we know this particular type of column is not available commercially, but it can be easily made by a professional glassblower. Flat sinters (No. 1 or 2) are preferred, but usually cannot be blown above 8 or 9 cm diameter. When using a larger diameter column,

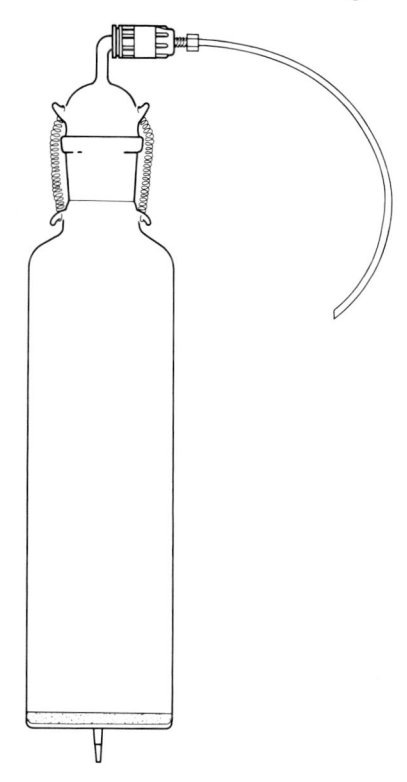

Figure 1. Glass column used for short column chromatography.

Appendix I

which will invariably have a slightly domed sinter, it will be necessary to slurry pack a layer of acid-washed sand in the bottom of the column to obtain a flat surface. However, the packing of the Kieselgel must then be carried out *very* carefully so as not to disturb the layer of sand, and this is quite difficult to achieve. Since chromatography in small columns is quite fast, it is better for the less experienced to duplicate a smaller scale separation in a column with a flat sinter rather than attempt a single large-scale separation in a column with a domed sinter.

Use about 15 – 30 g of Kieselgel 60H per gram of crude material to be purified depending upon the difficulty of the separation. In the purification of deoxyribonucleotide monomers used in the phosphotriester method (Chapter 4) where the silica gel t.l.c.s are uncomplicated, use 15 g of adsorbent per gram of crude material. Aim to use a column of dimensions such that the height of the packed bed of silica is about the same as the bed width or a little less [N.B. 100 g of Kieselgel (Merck Art 7736) packed in dichloromethane occupies a volume of ~300 ml; thus 130 g of silica gives a bed of ~8.5 cm deep x 7.5 cm wide]. In particular there should be at least one bed volume of space for eluent left in the packed column in order to avoid too frequent topping up with solvent.

The column set up and running, illustrated by the purification of about 9 g of crude monomer is as follows.

(i) Clamp the column (~7.5 cm internal diameter) vertically next to a u.v. detector and fraction collector and connect the column Luer outlet with narrow bore tubing to the fraction collector *via* the u.v. cell.

(ii) Pour about 100 ml of dichloromethane/1% triethylamine (use 0.5% pyridine in place of triethylamine when chromatographing protected deoxyribonucleosides) into the column, fit the gas inlet cone (the nitrogen inlet line should have an on/off valve and a pressure release valve between the on/off valve and the cone) into the column top (this prevents the column top being blown out by nitrogen pressure).

(iii) Open the column outlet and apply about 2.5 p.s.i. of nitrogen pressure (use a needle valve regulator) to the column (remember to shut the pressure release valve and open the on/off valve) so that air bubbles trapped in the glass sinter are removed.

(iv) Then, while there is still liquid covering the sinter, open the pressure release valve, shut the on/off valve, clamp the column outlet tube and remove the gas inlet cone.

(v) Suspend about 130 g of Kieselgel 60H (sufficient adsorbent to purify the crude monomer from a 10 mmol scale preparation) in about 600 ml of dichloromethane/1% triethylamine and stir the slurry to remove any trapped air bubbles.

(vi) Pour all of the well-stirred slurry directly into the column and pack the adsorbent under 2.5 p.s.i. of nitrogen pressure (remember to fit gas inlet cone and spring clips, then shut pressure release valve, open the nitrogen on/off valve and open the column outlet). The Kieselgel will pack down to give a bed about 8.5 cm deep.

(vii) When the bed height has stabilised shut the on/off valve, open the pressure release valve and clamp the column outlet tube. Then, with the aid of a glass funnel, carefully apply fine acid-washed sand evenly on to the surface of the Kieselgel to give a layer about 0.5 cm deep. This operation should only be carried out with several centimetres of liquid above the bed surface so that disturbance to the bed surface is minimal.

(viii) Dissolve the material to be purified (should be reasonably free of pyridine, otherwise evaporate dry toluene from the sample *in vacuo* to achieve this) in about 25 – 30 ml of dichloromethane/1% triethylamine.

(ix) Drain the liquid level in the column down to the surface of the sand and, with the column outlet closed, apply the sample solution slowly and evenly, using a 25 ml pipette on to the top of the column bed (keep the pipette tip ~1 cm above the surface of the material in the column to minimise any disturbance). The careful application of the sample in an even band is of the *utmost importance* in ensuring good resolution.

(x) Run the sample solution into the column bed under nitrogen pressure such that the liquid level reaches the surface of the sand.

(xi) Next add 3 x 15 ml portions of dichloromethane/1% triethylamine and run each of these into the column bed so that all of the sample is washed on to the Kieselgel.

(xii) Elute the column with about 200 ml of 4% ethanol in dichloromethane/1% triethylamine followed by 1 – 1.5 litres of 7% ethanol in dichloromethane/1% triethylamine under 2.5 p.s.i. of nitrogen, taking great care not to disturb the top of the bed when topping up the column with eluent.

(xiii) Collect fractions of about 20 – 25 ml, monitor these by $A_{280\,nm}$ on the flow cell u.v. detector and also by silica gel t.l.c. in ethanol/chloroform (1:9 v/v) containing 0.5% pyridine.

(xiv) Pool only those fractions containing pure product as determined by silica gel t.l.c., and then evaporate the pooled fractions to dryness *in vacuo*.

2. THIN LAYER CHROMATOGRAPHY ON SILICA GEL

For t.l.c. of the various protected deoxyribonucleosides and deoxyribonucleotides we recommend the use of silica gel $60F_{254}$ (0.2 mm thick layer) coated on aluminium sheets [20 x 20 cm (Merck, Art 5554)]. The advantage of these sheets is that they can easily be cut to any size needed. The most useful solvent systems are ethanol/chloroform (1:9 v/v) containing 0.5% pyridine and ethyl acetate/acetone/water (5:10:1 by vol.) containing 0.5% pyridine. Pyridine is added to prevent loss of acid-labile 5' protecting groups.

(i) Pour enough of the running solvent into a suitable glass tank (or beaker without a lip, or jam jar) equipped with a *tight fitting cover*, such that the liquid depth in the tank is about 0.5 cm. Cut a suitably sized t.l.c. plate (e.g., 4 – 5 cm wide x 10 cm long) from a 20 x 20 cm plate (the silica gel side can be lightly marked out in pencil without damaging the layer) using scissors or a scalpel and then mark a baseline about 1 cm in from the bottom edge of the plate.

Appendix I

(ii) Apply the t.l.c. sample in solution (1 – 2 μl) using a glass micro-capillary to give a tight spot on the baseline (keep the sample spot at least 1.5 cm away from the edge of the plate). If you apply a sample in pyridine or some other high boiling solvent blow most of the solvent off the sample spot using a cool air stream (a hairdryer is very useful).

(iii) Stand the plate vertically in the t.l.c. tank, put the cover on and run the plate in the ascending mode until the solvent front almost reaches the top of the plate.

(iv) Remove the plate from the tank, mark the solvent front and then dry the plate thoroughly.

(v) Visualise the plate under a short wavelength u.v. lamp (**WARNING: use eye protection**); nucleoside and nucleotide derivatives and other compounds absorbing in the u.v. at about 260 nm will appear as dark purple spots on a pale green fluorescent background.

An assortment of specific sprays can then be used to detect the presence of particular chemical groups. One of the most useful of these is a perchloric acid/ethanol spray. Spray the plate with 60% aqueous perchloric acid/ethanol (3:2 v/v) using an aerosol sprayer in a fume hood (in some laboratories 20% sulphuric acid is used instead); compounds bearing dimethoxytrityl groups appear as orange spots and those bearing pixyl protecting groups appear as yellow spots. On heating the plate at 100°C in an oven for a few minutes the spots will become orange-brown and green-yellow, respectively. Moreover sugar derivatives, e.g., base protected nucleosides bearing free OH groups and no DMTr or pixyl groups, appear as black spots after spraying and heating as above.

It is useful to measure the R_f of a compound on t.l.c. since it is characteristic of that compound in a particular solvent system. The R_f is the distance the compound has moved from the baseline divided by the distance travelled by the solvent front from the baseline. If two compounds have identical R_fs in several different solvent sytems (the R_f in each will usually be different) and moreover if they run as single spots in several systems when they are mixed, it is probable that they are the same. Note that the R_f varies to some extent with loading (a badly streaking or tailing spot usually indicates sample overloading) and also with the distance that the plate is run. It is therefore a good idea to standardise and run all plates for, say, 9 – 10 cm, and to apply roughly similar quantities of material to obtain reasonably consistent t.l.c. results.

3. DISTILLATION

3.1 **Atmospheric Pressure**

(i) Pour the liquid to be distilled into an appropriately sized one-neck, Quickfit, round-bottomed flask (fill to about two-thirds full only), add any drying agent required (see Chapter 4, Section 3.3), plus a few anti-bumping granules.

(ii) Stand the flask in an isomantle of the correct size and fit a reflux condenser into the Quickfit socket.

Appendix I

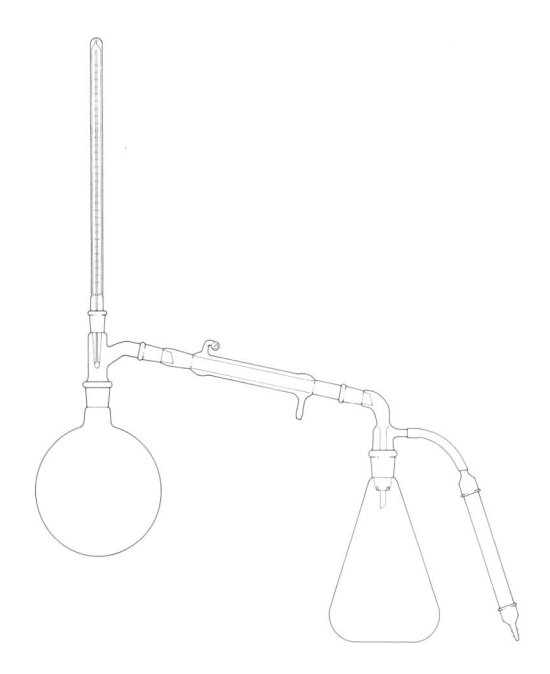

Figure 2. Apparatus for distillation under atmospheric pressure.

(iii) Fit a drying tube containing self-indicating silica gel or granulated calcium chloride into the top of the condenser and support the apparatus with a retort stand and clamps.
(iv) Turn on the isomantle and the condenser water (remember that the water inlet should be lower than the outlet) and when the liquid starts to boil adjust the isomantle control so that the liquid refluxes gently.
(v) At the end of the reflux period turn off the heating and arrange the apparatus for distillation as shown in *Figure 2*. Remember to fit the receiver adaptor with a drying tube.
(vi) Set the isomantle controller such that the liquid distils over at a reasonable rate.
(vii) Discard the first few percent of the distillate and collect the main fraction (distillation temperature should not fluctuate) in a Quickfit round-bottomed or conical flask.
(viii) Stop the distillation *well before* the distillation flask goes dry since some thermal decomposition may occur on the walls.

To carry out a distillation under nitrogen, use a two-necked distillation flask and pass a slow stream of dry nitrogen through the apparatus *via* the second neck of the flask. When distilling pyridine, use a long Vigreux column between the distillation flask and the still-head.

3.2 Reduced Pressure

The apparatus for a vacuum distillation is shown in *Figure 3*. It can be built large-

Appendix I

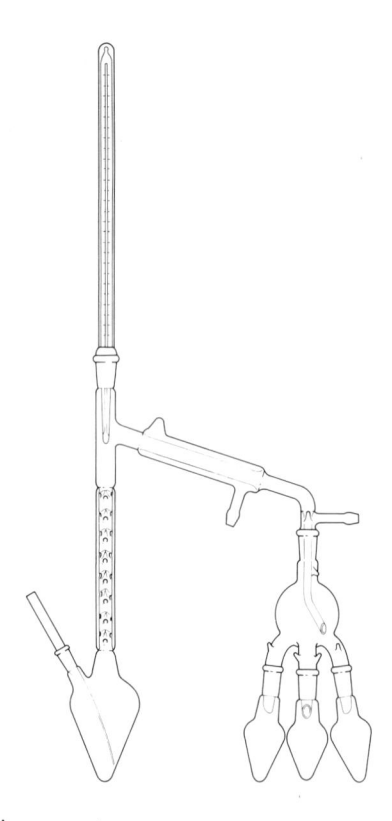

Figure 3. Vacuum distillation apparatus.

ly in one piece (this reduces to a minimum the number of joints that can leak) which comprises a two-necked flask (pear-shaped), vacuum jacketed Vigreux column, still-head, water condenser and a receiver adapter.

(i) Clamp the apparatus onto a retort stand.
(ii) Lightly grease all the joints, then fit a multi-position receiver adapter (a 'cow') plus collection flasks (these should be wired on) onto the receiver adapter and fit a thermometer in the still-head.
(iii) Half fill the pear-shaped flask with the liquid to be distilled and then fit a capillary bleed (a fine drawn out glass capillary reaching to the bottom of the flask) into the side neck so that dry air or nitrogen can be bled in during the distillation. A suitable capillary can be pulled with a little practice from Quickfit item MF 15/0, using a gas/oxygen torch. If the capillary can be easily bent without breaking then it is fine enough to be used.
(iv) Attach vacuum tubing from either an oil pump or a water pump (depending upon the vacuum required) to the conection on the receiver adapter, evacuate the system and adjust the dry air or nitrogen flow through the capillary bleed as necessary.
(v) Heat the flask in a magnetically stirred, thermostatistically controlled oil bath and when the liquid starts to boil endeavour to maintain the oil bath temperature at about 20°C higher than the distillation temperature.

(vi) Collect the forerun in one of the collection flasks and, when the distillation temperature is steady, rotate the 'cow' to a clean collection flask and collect the fraction boiling at the correct temperature.
(vii) At the end of the distillation raise the apparatus clear of the oil bath and then release the vacuum.

When distilling DMF, use a long glass column filled with glass helices in place of the Vigreux column.

4. NINHYDRIN TEST FOR PRIMARY AMINES

Prepare the following ninhydrin test solutions (Kaiser,E. *et al.*, 1970, *Anal. Biochem.*, **34**, 595-598):

(i) 2 ml of 0.1 M aqueous potassium cyanide diluted to 100 ml with pyridine;
(ii) 500 mg of ninhydrin in 10 ml of ethanol;
(iii) 80 g of phenol in 20 ml of ethanol.

Add one drop of each of the three test solutions to the test sample (a few beads of thoroughly washed support or one drop of liquid) in a small glass vial. Heat the vial in an oven at 100°C for 5 min. A blue or mauve colour on the beads and in solution indicates the presence of free primary amino groups. A blank (one drop of each solution) gives a pale yellow colour normally.

The test should also be made as a check for the presence of ammonia or primary amine contaminants in solvents such as pyridine and DMF.

APPENDIX II

H.p.l.c. Column Packing Techniques

LARRY W. McLAUGHLIN and NORBERT PIEL

The following procedure allows one to produce self-packed h.p.l.c. columns of high quality at a reasonable price. The columns are slurry-packed in methanol followed by a wash with water. An h.p.l.c. pump which can generate a solvent flow of 10 ml/min at 200 bar is adequate. Column quality and life are slightly better if a constant pressure, packing pump is used, which can produce a flow of 100 ml/min and pressure in excess of 400 bar.

1. EQUIPMENT

The following hardware includes the column itself. High quality stainless steel tubing can be purchased from a number of companies. The part numbers listed correspond to those from *Swagelok*. Other stainless steel compression fittings can also be used but it is important not to mix fittings from different companies since the angles and tolerances will vary.

Packing column:
        stainless steel 9.5 mm (3/8 in) o.d., 5.7 mm i.d., 700 mm long
        column end fitting 3/8 in − 1/16 in SS-600-6-1ZV
        reducing union 3/8 in − 1/4 in SS-600-604.

Analytical h.p.l.c. column:
        stainless steel 6.4 mm (1/4 in) o.d., 4.6 mm, i.d., 250 mm
        column end fittings 1/4 in − 1/16 in SS-400-6-1ZV-SA (5 μm frit)
                                                       SS-400-6-1ZV-S5 (2 μm frit)

Pump: H.p.l.c. or column packing pump (see Shandon Southern Inc.) and 2 m of
        1/16 in o.d. stainless steel tubing.

Assembly of the apparatus is as diagrammed in *Figure 1*. The packing column can be of other dimensions but should have an internal volume of at least 18 ml.

2. PACKING PROCEDURE

The column will be initially packed upwards as indicated by the relative positions of the columns in *Figure 1*. This is done for two reasons. It ensures that any air in either the packing or chromatographic column escapes and any large aggregates or dense contaminating material should sink to the bottom of the packing column and not be incorporated into the chromatographic column.

(i)   Assemble the apparatus as shown in *Figure 1* allowing enough 1/16 in stainless steel tubing between the pump and the packing column, and bet-

Appendix II

[Figure 1: Column packing apparatus, showing waste bottle, chromatography column with column end fittings (¼ in – 1/16 in, 2 μm frit and 5 μm frit), reducing union (3/8 in – ¼ in), packing column, column end fitting (3/8 in – 1/16 in), pump, and solvent bottles (CH₃OH and H₂O).]

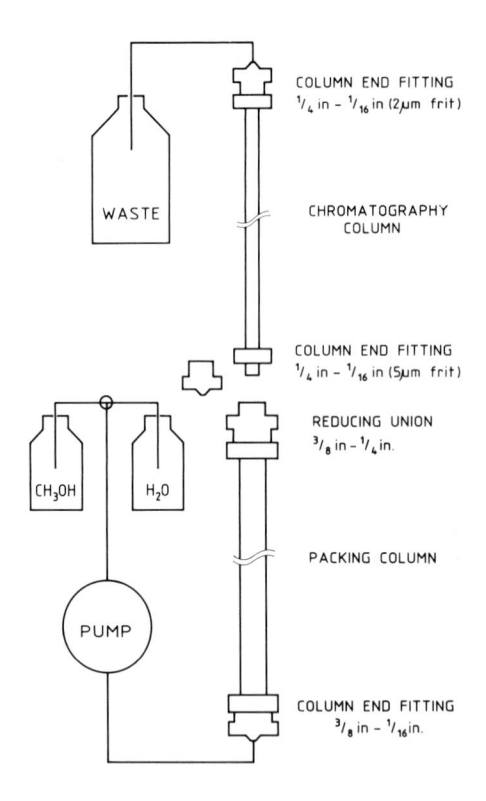

ween the chromatographic column and the waste bottle that the two columns when coupled together can additionally be turned 180°.

(ii) To a small screw-top bottle (glass scintillation vials do nicely) with a minimum 20 ml volume add 3.3 g of the desired 5 μm column support and 15 ml of methanol. (**Caution: manufacturers of microparticulate silicas recommend that dry material is weighed out in a fume hood since the danger of inhaled microparticulates of this type is at present unknown.**)
(iii) Screw the bottle closed and shake to suspend the silica support.
(iv) To ensure suspension and the breakdown of large aggregate particles place the bottle in an ultrasonic bath for 2 min (Caution: too long in the ultrasonic bath may result in fracturing of the silica support).
(v) Charge the pump with methanol and assemble the column apparatus as shown in *Figure 1*, such that the only connection which is still open is that between the chromatographic column and the packing column.
(vi) Add the silica suspension with a disposable pipette to the packing column.
(vii) Connect the chromatography column and the packing column together and start the pump (**wear safety glasses!**). It is recommended that the time between addition of the suspension to the packing column and starting of the pump be kept to a minimum.

(viii) Pack the column at 400 bar (constant pressure, packing pump) for 15 min or at 200 bar (h.p.l.c. pump 10 ml/min initial flow) for 30 min.
(ix) Rotate the joined columns 180° and continue packing with water for 15 min or 30 min, respectively. Then stop the pump.
(x) After the pressure has stabilised at 0 bar, loosen the compression fitting between the chromatography column and packing column.
(xi) Using a spanner (wrench) on the outflow fitting of the chromatography column, rotate the entire column without removing it from the reducing union attached to the packing column. This ensures that the surface of the column support at the end of the chromatography column is cleanly broken off from excess material in the packing column.
(xii) Remove the chromatography column from the reducing union, repair the surface of the support with a spatula and using excess column support from the packing column if necessary, and clean the outside of the compression fitting and attach the column end fitting to the chromatography column.
(xiii) Mark the direction of flow and/or inlet and outlet of the column. The column is immediately useable.

Larger h.p.l.c. columns (9.4 x 250 mm) can be packed using the same procedure with the following exceptions: use a packing pump with an initial solvent flow of approximately 100 ml/min at a pressure of 300 bar and a stainless steel packing column with a 100 ml internal volume. For a 9.4 x 250 mm column suspend 16 g of support in 90 ml of methanol and pack the column at 300 bar for 30 min with methanol following with water for 30 min.

3. REPACKING THE SAME COLUMN SUPPORT

In our experience with reversed-phase columns (ODS-Hypersil), poor resolution, broad peaks or double peaks are generally related to disruption of the column bed or irreversibly bound material. In general, a large portion of the column support from a column which exhibits poor resolution is in good condition. To take advantage of this, particularly in the case of larger columns, and to extend the life of the column support, use the following procedure.

(i) Wash the column with water and then methanol.
(ii) Open the inlet column fitting and attach the outlet fitting to a pump charged with methanol.
(iii) Increase the pump flow and/or pressure slowly until the column support is slowly extruded from the column.
(iv) Remove the material which is coloured (usually varying from yellow to brown) from the end of the column (~1 cm).
(v) Collect the remainder of the column support in a glass screw-top bottle.
(vi) Add enough new support to make up for that discarded, plus a slight excess (generally a total of 0.5 g for analytical columns and 2.0 g for preparative columns).
(vii) Treat this materal as described above and repack into the column.

Appendix II

4. TROUBLESHOOTING

One of the most common problems occurring with h.p.l.c. columns is high column back pressures. This is almost always a result of a clogged column inlet frit. The inlet frit can be removed without disturbing the column bed and replaced with a new one. Clogged frits can generally be cleared by immersion in 30% aqueous nitric acid overnight.

On rare occasions high pressures result from clogged 1/16 in tubing. This can generally be confirmed by removing the column and observing that the pressure remains high. We have only had difficulty with clogged tubing between the injection port and the column inlet and on one occasion between the column outlet and the detector.

APPENDIX III

Suppliers of Chemicals and Equipment for Oligonucleotide Synthesis

Many of the larger companies have subsidiaries in other countries whilst most of the smaller companies market their products through agents. The primary addresses of the companies listed here are the head offices. Second addresses refer usually to subsidiaries or local agents in the USA (for European companies) or in the UK (for USA companies). Telephone numbers have also been given but country and area codes may not be the same for all locations.

Companies marketing equipment for DNA synthesis (manual or automated 'gene machines') are denoted by*.

Aldrich Chemical Co., P.O. Box 355, Milwaukee, WI 53201, USA (tel. (414) 273 3850) and The Old Brickyard, New Road, Gillingham, Dorset SP8 4BR, UK (tel. (07476) 2211)

American Bionuclear, 4560 Horton Street, Emeryville, CA 94608, USA (tel. (415) 652 4466) and Genofit, 5, rue des Falaises, 1205 Geneva, Switzerland (tel. (022) 298430)

Applied Biosystems Inc.*, 850 Lincoln Centre Drive, Foster City, CA 94404, USA (tel. (415) 570 6667) and Applied Biosystems Ltd., Birchwood Science Park, Warrington, Cheshire, UK (tel. (0925) 825650)

Bachemgentec Inc.*, 3136 Kashiwa Street, Torrance, CA 90505, USA (tel. (213) 539 4171) and Bachem UK, 69 High Street, Saffron Walden, Essex CB10 1AA, UK (tel. (0799) 26465)

Beckman Instruments Inc.*, P.O. Box 10200, Palo Alto, CA 943054, USA (tel. (415) 857 1150) and Beckman RIIC Ltd., Progress Road, Sands Industrial Estate, High Wycombe, Bucks. HP12 4SL, UK (tel. (0494) 41181)

Bethesda Research Laboratories Inc., P.O. Box 6009, Gaithersburg, MD 20877, USA (tel. (301) 840 8000) and BRL UK Ltd., P.O. Box 145, Science Park, Milton Road, Cambridge CB4 4BE, UK (tel. (0223) 315504)

Biosearch*, 2980 Kenner Boulevard, San Rafael, CA 94901, USA (tel. (415) 459 3907/(800) 227 2624) and New Brunswick Scientific UK, Ltd., 26-34 Emerald Street, London WC1N 3QA, UK (tel. 01-404 4515)

Biosyntech*, Stresemannstrasse 268-280, D-2000 Hamburg 50, FRG (tel. 40-8500051)

Boehringer Mannheim GmbH, Biochemica, P.O. Box 310120, D-6800 Mannheim 31, FRG (tel. (UK) (07916) 71611) and Boehringer Mannheim

Appendix III

Biochemicals, 7941 Castleway Drive, P.O. Box 50816, Indianapolis, IN 46250, USA (tel. (317) 849 9350)
Cambridge Research Biochemicals, Ltd.*, Button End Industrial Estate, Harston, Cambridge CB2 5ND, UK (tel. (0223) 871674) and 1887 Park Street, Atlantic Beach, NY 11509, USA (tel. (526) 239 3831)
ChemGenes Corp., 296 Newton Street, Waltham, MA 02154, USA (tel. (617) 894 9118)
Creative Biomolecules, 385 Oyster Point Boulevard, Suite 4, South San Francisco, CA 94080, USA (tel (415) 583 0844/(800) 447 7500)
Cruachan Chemical Co., Ltd. (Cruachem)*, 11 Napier Square, Livingstone EH54 5DG, UK (tel. (0506) 32146) and Cruachem Inc., P.O. Box 5787, Bend, OR 97708, USA (tel. (503) 382 5860)
Fluka AG, Chemische Fabrik, CH-9470, Buchs, Switzerland (tel. 085 602 75) and Fluorochem, Ltd., Peakdale Road, Glossop, Derbyshire SK13 9XE, UK (tel. (04574) 62518)
Genetic Design, Inc.*, 111 School Street, Watertown, MA 02172, USA (tel. (617) 923 1175)
Labor-Service GmbH*, Damlerweg 2, D-6100 Darmstadt, FRG (tel. 06151 82851)
Merck,E., Frankfurter Strasse 250, D-6100 Darmstadt, FRG (tel. Darmstadt 06151) and E.M. Industries Inc., 480 Democrat Road, Gibbstown, NJ 08027, USA (tel. (609) 423 6300)
New England Biolabs, 283 Cabot Street, Beverly, MA 01915, USA (tel. (617) 927 8576)
Omnifit, Ltd.*, 51 Norfolk Street, Cambridge CB1 2LE, UK (tel. (0223) 69841) and Omnifit, Inc., 1887 Park Street, Atlantic Beach, NY 11509, USA (tel. (506) 239 1655)
Peninsula Laboratories, Inc., 611 Taylor Way, Belmont, CA 94002, USA (tel. (415) 592 5392/(800) 922 1516) and Merseyside Laboratories, P.O. Box 62, 17K Westside Industrial Estate, Jackson Street, St. Helens, Merseyside WA9 3AJ, UK (tel. (0744) 612108)
Pharmacia P-L Biochemicals Inc., 800 Centennial Avenue, Piscataway, NJ 08856, USA (tel. (201) 457 8000) and Pharmacia Ltd., Pharmacia House, Midsummer Boulevard, Milton Keynes MK9 3HP, UK (tel. (0908) 661101)
Sigma Chemical Company, P.O. Box 14508, St. Louis, MO 63178, USA (tel. (314) 771 5750/(800) 325 3010) and Sigma London Chem. Co., Ltd., Fancy Road, Poole, Dorset BH17 6NH, UK (tel. (0202) 733114)
Systec, Inc.*, 3816 Chandler Drive, Minneapolis, MN 55421, USA (tel. (612) 788 9701)
Vega Biochemicals*, P.O. Box 11648, Tucson, AZ 85734, USA (tel. (602) 746 1401/(800) 528 4882) and C.P. Laboratories, Ltd., P.O. Box 22, Bishop's Stortford, Herts, UK (tel. (0279) 53734)
Yuki Gosei Kogyo Co., Ltd., 17-4 2-Chome Kyobashi Chuo-Ku, Tokyo, Japan (tel. 03 (567) 5481) and Toyo Menka Kaisha, Ltd., St. Alphage House, 2 Fore Street, London EC2Y 5DQ, UK (tel. 01-628 2591)

Appendix III

General Chemical and Equipment Companies

Ace Glass Inc., P.O. Box 688, 1430 Northwest Boulevard, Vineland, NJ 08360, USA

Analtech Inc., 75 Blue Hen Drive, P.O. Box 7558, Newark, DE 19711, USA

J.T. Baker Chemical Company, Phillipsburg, NJ 08865, USA

BDH Chemicals Ltd., Poole BH12 4NN, Dorset, UK

Becton-Dickinson and Co., Rutherford 1, NJ 07070, USA

Fisher Scientific Co., Fairlawn, NJ 07410, USA

Hamilton Co., Reno, NJ, USA

Hopkin and Williams, P.O. Box 1, Romford, Essex RM1 1HA, UK

Janssen Life Sciences Products, Turnhoutsweg 30, B-2340 Beerse, Belgium

Mallinckrodt, Inc., Paris, KY 40361, USA

Millipore Corporation, Bedford, MA 01730, USA, or Millipore House, Abbey Road, London NW10 75P, UK

Nalge Company, Division of Sybron Corporation, Rochester, NY 14602, USA

Pierce Chemical Company, P.O. Box 117, Rockford, IL 61105, USA

Pierce and Warriner (UK) Ltd., 44 Upper Northgate Street, Chester, Cheshire CH1 4EF, UK

Savant Instruments, Hicksville, NY 11801, USA

Schleicher and Schuell Inc., Keene, NH 03431, USA

INDEX

Acids and bases,
  impurities, 19
Acylation,
  DNA, 11
  N-, 24-27
Alkylation,
  DNA, 11
Analysis of oligodeoxyribonucleotides,
  total digestion, 126
Analysis of oligoribonucleotides,
  total digestion, 128-131
Apolipoprotein E, 5
Apparatus,
  synthesis, 19-21, 54-57, 88, 101-103
Assembly cycle,
  procedure, 49-51, 58-67, 87-89,
    103-106
Assembly,
  oligoribonucleotides, 174-177

Bond-Elute,
  see Sep-Pak,

Capping, 16, 50, 52
cDNA cloning, 5
Chain assembly,
  DNA, 15-16
Chloro-N,N-diisopropylamino
  methoxyphosphine, 41
Chromatography,
  short column, 199-201
Cleavage,
  linkage, 67, 108-109
Controlled pore glass, 49, 85, 91, 103
Coupling reaction, 15, 50
Crystallisation,
  oligonucleotides, 8

Denaturing agents, 9
Deoxyadenosine, N-6 protection,
  27, 92
Deprotection,
  acidic, 16, 49-50, 77
  final, 17, 67-70, 89, 108-109, 156,
    177-178
Desalting, 109
Dichloro-1,1,3,3-tetraisopropyl-
  disiloxane,
    preparation, 159-160
Dideoxy sequencing, 27-28
Dimethoxytrityl,
  assay, 48, 91
  group, 12, 23, 49-50

Dimethylaminopyridine, 51,52
Distillation, see solvents, distillation
DNA,
  chemical reactivity, 10-11
  duplex, 1, 9-10
  Z-form, 7
DNA ligase, T4, 2
DNA structure,
  primary, 8-9
  secondary, 8-10
DNA synthesis machines, 21, 77, 135,
  211-213

Electrophoresis,
  cellulose acetate, 141-142

Fractosil support, 46-47

Gel electrophoresis, see polyacrylamide
  gel electrophoresis
Gene synthesis, 1-3
Guanine,
  O-6 modification, 89
  O-6 protection, 24, 29-33, 154
    168-169

Heterocyclic bases, 11, 12
Hexadecyloxy-4′,4″-dimethoxytrityl,
  ribonucleoside monomers, 171-174
H.p.l.c., 117-133
  column packing, 207-210
  column supports, 119
  equipment, 117-119
  ion exchange, 109, 120-122
  isolation of oligonucleotides, 124,
    127-130
  mixed-mode, 131-132
  reversed-phase, 110, 124, 191-192
Hybridisation, 9
Hydrogen bonding, 8

Internucleotide,
  cleavage, 13
  linkage, 11, 36

Labelling,
  5′-, 139-140
  3′-, 140
Large-scale synthesis, see scale up
Ligation, 2-3
Linkers, 3

M13, 16
Machines, DNA, synthesis, see DNA
  synthesis machines

Index

Maxam-Gilbert sequencing,
  see sequence analysis
Melting temperature, 5, 6, 9-10
Metal ions, 19
Methyl phosphodichloridite, 39-40
Mitsunobu alkylation, 32-33
Mobility shift analysis, *see* sequence analysis,
Monomers,
  deoxyribonucleotide, 41-45, 91-96
MSNT,
  preparation, 98
Mutagenesis,
  site-directed, 6, 7

Ninhydrin test, 205
N.m.r. spectra, 8, 135

Oligoribonucleotide synthesis,
  chemical, 153-184
  enzymatic, 185-198
Oxidation,
  nucleic acids, 50

Parallel synthesis,
  see simultaneous synthesis
Per-acylation, 24-25
Phosphite-triester synthesis, 16, 35-81
Phosphitylation agent, 40-41
Phosphodiester synthesis, 13
Phosphoramidites,
  deoxyribonucleoside, 35-44
Phosphorylating agent, 94, 174-175
Phosphotriester synthesis, 16, 83-115
Pixyl group, 12, 93, 94-96
Polyacrylamide gel,
  electrophoresis, 70-71, 138-139
  preparation, 70, 147
Polydimethylacrylamide-Kieselguhr, 15, 85, 89
Polynucleotide kinase T4, 2, 139, 194, 195
Polynucleotide phosphorylase,
  Micrococcus luteus, 186, 192-194
Primers,
  mismatched, 6
  oligodeoxyribonucleotide, 3-5
Probes,
  hybridisation, 5-6, 7, 10
  mixed-sequence, 5-6, 75, 150-151
Protecting groups, 12-13, 23-24, 86-87

Protection,
  one-flask, 28-29
  strategy, 23
  transient, 25-27, 33
Purification,
  oligonucleotides, 70-72, 117-133, 191-192

Reagents,
  purification, 51-52, 98-101
Reflux,
  continuous, 53-54
RNA ligase T4,
129-130, 188-189, 194, 195-196
RNases, 190-191

Scale up, 72-73, 107-108
Sep-Pak, 71-72, 191-192
Sequence analysis,
  chemical, 136, 143-147
  enzymatic, 135, 141-143
Sequencing gel, *see* polyacrylamide gel preparation
Simultaneous synthesis, 73-75, 86
Snake venom phosphodiesterase, 141
Solid-phase synthesis,
  method, 14, 34
Solvents,
  distillation, 53, 202-205
  purification, 18, 53-54, 98-101
  purity, 18, 51-52, 84
Structural studies,
  DNA, 7-8
Succinate linkage, 15, 85
Succinates,
  preparation, 47-48, 90, 96-98
Sulphonylation/displacement, 31-32
Supports,
  preparation, 45-49, 89-91
  solid-phase, 14-15, 45, 84-86
Symmetrical anhydride, 90-91

Terminal transferase, 140
Tetrahydropyranylribonucleoside monomers,
  synthesis, 156-171
Thin layer chromatography,
  DEAE cellulose, 142
  silica gel, 137, 201-202
Triphenylmethyl group, 13
Trityl assay, *see* dimethoxytrityl assay
Trityl cations,
  coloured, 75-76

Index

Trityl derivative,
　lipophilic, 153
Trouble shooting, 20, 21, 83-84, 106, 210

Unblocking, *see* deprotection, final

Wallace rule, 10
Water,
　in solvents, 19

Yield,
　repetitive, 17-18, 110

Zinc bromide, 16, 23

Forthcoming

the Practical Approach series

Nucleic acid hybridisation
a practical approach
Edited by B D Hames and S J Higgins
A practical laboratory-bench manual of techniques for identifying and analysing the structure of specific gene sequences. This book is unique in bringing together the techniques' major applications at both the theoretical and the practical levels.
Due September 1985; 250pp (approx); 0 947946 23 3 (softbound)

Animal cell culture
a practical approach
Edited by R I Freshney
After an introductory chapter dealing with basic techniques, this book provides detailed protocols both for traditional areas like organ culture, characterisation and storage, and for those in developing fields. These include serum-free media, cell separation and *in situ* hybridisation.
Due October 1985; 250pp (approx); 0 947946 33 0 (softbound)

Photosynthetic energy transduction
a practical approach
Edited by M F Hipkins and N R Baker
An up-to-date laboratory manual for researchers and students wishing to learn a wide range of techniques for the study of photosynthetic energy transduction.
Due late 1985; 250pp (approx); 0 947946 51 9 (softbound)

Biochemical toxicology
a practical approach
Edited by K Snell and B Mullock
Chapters written by laboratory experts provide practical guidance and 'tricks of the trade' for the most useful techniques in toxicological research. The book is unique as a guide for researchers at all levels, especially those in pharmaceutical and agrochemical laboratories.
Due late 1985; 250pp (approx); 0 947946 52 7 (softbound)

PRICES TO BE ANNOUNCED

◇ IRL PRESS

IRL Press Ltd, PO Box 1, Eynsham, Oxford OX8 1JJ, UK
IRL Press Inc, Suite 907, 1911 Jefferson Davis Highway, Arlington, VA 22202, USA